Spring Boot
整合开发案例实战

颜井赞 编著

清华大学出版社
北京

内 容 简 介

随着Java Web项目的发展，各种开发框架与组件层出不穷，项目的配置越来越烦琐，项目部署也需耗费大量时间，给开发人员带来了诸多不便。Spring Boot的出现将开发人员从烦琐的项目配置中解放出来，让开发人员更专注于业务的实现，提高了开发效率。本书从Spring Boot项目实战的角度出发讲解Spring Boot的原理与整合使用，包括每个项目所使用的技术与编码实现过程。本书配套示例项目源码和PPT课件。

本书分为7章。第1章主要介绍Spring Boot的特性、核心模块、开发环境配置与简单使用，同时对Spring Boot源码进行简单分析；第2~7章主要介绍实战项目（包括员工管理系统、二手房管理系统、购物车管理、用户权限管理系统、小程序上报用户信息、模拟聊天室），每个典型的实战项目，都遵循项目的需求、设计以及实现流程，讲解项目的开发流程、使用技术和实现方法，同时介绍所用新工具的安装和使用，并对每一个项目做出简单总结。

本书内容由浅到深、解析详细、示例丰富，从实战角度指导读者使用Spring Boot进行项目开发，适合Spring Boot初学者快速入门以及具有一定经验的开发者提高技术整合能力，同时也适合作为高等院校相关专业的教材。

本书封面贴有清华大学出版社防伪标签，无标签者不得销售。
版权所有，侵权必究。举报：010-62782989，beiqinquan@tup.tsinghua.edu.cn。

图书在版编目（CIP）数据

Spring Boot整合开发案例实战 / 颜井赞编著. —北京：清华大学出版社，2023.1（2024.8 重印）
ISBN 978-7-302-62404-2

Ⅰ. ①S… Ⅱ. ①颜… Ⅲ. ①JAVA语言—程序设计 Ⅳ. ①TP312.8

中国国家版本馆CIP数据核字（2023）第016300号

责任编辑：夏毓彦
封面设计：王 翔
责任校对：闫秀华
责任印制：宋 林

出版发行：清华大学出版社
网　　址：https://www.tup.com.cn，https://www.wqxuetang.com
地　　址：北京清华大学学研大厦A座　　邮　编：100084
社 总 机：010-83470000　　邮　购：010-62786544
投稿与读者服务：010-62776969，c-service@tup.tsinghua.edu.cn
质量反馈：010-62772015，zhiliang@tup.tsinghua.edu.cn

印 装 者：三河市东方印刷有限公司
经　　销：全国新华书店
开　　本：190mm×260mm　　印　张：19　　字　数：513千字
版　　次：2023年3月第1版　　印　次：2024年8月第2次印刷
定　　价：89.00元

产品编号：081015-01

前　言

为什么要写这本书

随着互联网的飞速发展，各种基于互联网的系统深入我们的工作、学习、生活等方面。我们正在大力发展的智慧政务、智能家居、智慧城市等，都离不开专业开发技术的支持。目前 Java Web 开发的项目在各行各业都有广泛的应用。Java Web 项目的开发会基于各种各样的框架，目前较流行的是 Spring 框架，其他框架和组件也都做了适应和扩展。一个项目的实现，特别是复杂的大型综合性项目，所用到的框架更是数不胜数。而框架在融合的过程中会遇到各种各样的配置问题，例如版本不兼容、版本冲突、依赖缺失等，所以在 Java Web 项目开发过程中，框架整合显得越来越重要，也越来越烦琐。有鉴于此，Spring 团队设计并实现了 Spring Boot 脚手架。它相当于一个开发容器，我们可以按需拉取框架和组件，而不需要增加让人烦恼的配置文件或配置类。它会自动为我们的框架选择合适的配置并应用到容器。当然，我们也可以进行自定义配置。这个工具对开发者来说无疑是个福音，可以让千万开发者从烦琐的配置工作中解放出来，专注于业务逻辑的实现，使得开发效率得到大大的提高。

对一个 Java Web 开发人员来说，学习如何使用 Spring Boot 框架显得极为重要。这不仅能够提高自身的开发效率，更能增加系统的稳定性。本书就是基于此目的，从项目实战的角度为开发者讲解 Spring Boot 的框架组成、实现原理以及使用方法。通过一个个典型的项目应用，根据项目的侧重点来整合不同框架到项目中，利用每个框架技术的特点来实现项目业务逻辑，手把手地将项目的设计开发过程展示给读者。整本书的内容浅显易懂，讲解详细全面，是一本非常好的 Spring Boot 初学者的读物。

目前适用于初学者学习 Spring Boot 的书籍有许多，每一本书的侧重点不同。本书是在实战过程中逐步讲解技术知识点和应用场景，以及对应项目中的实践应用，将理论结合于实践，让读者加深理解，深入学习。对于初学者来说，如果想了解 Spring Boot 的技术原理，熟悉它的应用场景并能使用它实现相应的功能，这本书非常适用。对开发经验不够丰富的开发人员来说，本书案例整合了不同的技术，能提高他们在不同场景下采用不同技术来快速选型并实现项目的整合能力。

本书有何特色

（1）附带项目源代码，提高学习效率。为了便于读者理解本书内容，提高学习效率，笔者专门为本书的每一章内容都提供了项目源代码。

（2）涵盖 Spring Boot 开发 Web 项目的各种热门技术、主流框架、数据库及其整合使用。本书涵盖 Spring Boot、Spring、Spring MVC、Spring Data JPA、MyBatis、MyBatis Plus、Redis、MySQL、H2、PostgreSQL、Spring Security、Shrio、JWT、WebSocket、Thymeleaf 等主流框架的整合使用。

（3）涵盖多种前端 CSS、JS 框架。本书涵盖了多种涉及前端的框架的使用，包括 Vue、Node.js、LayUI、Element UI、uni-app 等。

（4）涉及多种开发工具的安装和使用。本书在前后端的开发过程中使用到了多种开发工具，演示了它们的安装过程和基本使用方法，例如 IntelliJ IDEA、HBuiderX、微信开发者工具等。

（5）项目案例驱动，应用性强。本书从第 2 章开始，每章都提供一个使用 Spring Boot 开发的项目典型案例，这些案例来源于笔者实际工作过程中所遇到的应用场景，具有较高的实战价值和参考性。这些案例都是在 Spring Boot 开发框架下根据业务特点整合了不同技术和框架，能够帮助读者在学习基础知识的同时，快速掌握以后开发中常用到的技术功能。读者在参考本书的同时，能够融会贯通地应用所学到的技术，将理论技术应用于实战开发中，快速提升理论与实战结合的开发经验。另外读者可以自行发挥思维，补充拓展，增加一些其他的功能，或者在原有代码基础上进行修改，便可用于实际的项目开发中。

（6）提供完善的技术支持和售后服务。本书提供了专门的技术支持邮箱，读者在阅读本书的过程中有任何疑问都可以通过该邮箱获得帮助。

本书内容及知识体系

第 1 章 Spring Boot 入门。本章主要介绍 Spring Boot 的特性、核心模块、开发环境配置和 Spring Boot 的使用，并对 Spring Boot 的源码进行分析。

第 2 章员工管理系统。本章主要介绍如何使用 Spring Boot 配置 MySQL 数据库、集成 MyBatis 插件，以及如何使用它们完成典型的员工管理系统，并使用 LayUI 框架搭建并实现配套的前端项目。

第 3 章二手房管理系统。本章主要介绍如何使用 Spring Boot 集成 Spring Data JPA，以及前端中 Node.js、Vue 和 Element UI 的介绍和使用。

第 4 章购物车管理。本章主要介绍如何使用 Spring Boot 集成 H2 数据库、Thymeleaf 模板和 Spring Security，介绍这些框架的概念、功能与使用方法，综合使用这些框架完成购物车的管理。

第 5 章用户权限管理系统。本章主要介绍如何使用 Spring Boot 集成 Shrio 框架、配置 PostgreSQL 数据库，以及如何使用 Shrio 完成权限校验。

第 6 章使用小程序上报用户信息。本章主要介绍如何使用 Spring Boot 集成 JWT、MyBatis-Plus 框架，如何使用 MyBatis-Plus 进行数据库操作，如何使用 JWT 进行登录认证，以及如何实现小程序登录。

第 7 章模拟聊天室。本章主要介绍如何使用 Spring Boot 集成 WebSocket，并做对应的配置，如何使用 WebSocket 协议进行前后端消息交互，实现模拟聊天室功能。

适合阅读本书的读者

- 需要学习 Java Web 开发技术的人员。
- 需要学习 Spring Boot 框架的使用方法的开发人员。
- 希望整合不同技术、提高项目开发水平的 Spring Boot 开发人员。
- 希望借鉴项目案例的开发人员。
- 软件开发项目经理。
- 专业培训机构的学员。
- 高等院校计算机相关专业的学生。

配套示例项目源码、PPT 课件下载

本书配套示例项目源码、PPT 课件，需要用微信扫描下面的二维码获取，可按扫描后的页面提示填写你的邮箱，把下载链接转发到邮箱中下载。如果下载有问题或阅读中发现问题，请用电子邮件联系 booksaga@163.com，邮件主题写"Spring Boot 整合开发案例实战"。

笔 者
2023 年 1 月

目 录

第 1 章 Spring Boot 入门 1
1.1 Spring Boot 介绍 1
1.1.1 Spring Boot 简介 1
1.1.2 Spring Boot 核心模块 3
1.2 开发环境配置 10
1.2.1 安装与配置 JDK 10
1.2.2 安装 IntelliJ IDEA 13
1.2.3 安装 Maven 14
1.2.4 安装 MySQL 15
1.3 使用 Spring Boot 19
1.3.1 快速搭建一个项目 19
1.3.2 运行发布项目 21
1.4 Spring Boot 源码分析 25
1.4.1 入口类@SpringBootApplication 25
1.4.2 深入理解自动配置 27
1.5 本章小结 29

第 2 章 员工管理系统 30
2.1 项目技术选型 30
2.1.1 MyBatis 30
2.1.2 框架搭建 32
2.2 项目前期准备 36
2.2.1 项目需求说明 36
2.2.2 系统功能设计 37
2.2.3 系统数据库设计 37
2.2.4 系统文件说明 38
2.3 项目前端设计 39
2.3.1 登录注册 39
2.3.2 部门管理 42
2.3.3 员工信息管理 44
2.3.4 工资管理 49
2.3.5 考勤记录管理 52
2.4 项目后端实现 53
2.4.1 通用分页类 53
2.4.2 通用返回结果 57
2.4.3 登录/注册 57
2.4.4 部门管理 61
2.4.5 员工信息管理 63
2.4.6 工资管理 64

  2.4.7 考勤记录管理 ······ 65
 2.5 项目总结 ······ 68

第 3 章　二手房管理系统 ······ 69

 3.1 项目技术选型 ······ 69
  3.1.1 Spring Data JPA ······ 69
  3.1.2 Node.js ······ 73
  3.1.3 Vue 和 Element UI ······ 75
  3.1.4 框架搭建 ······ 76
 3.2 项目前期准备 ······ 79
  3.2.1 项目需求说明 ······ 79
  3.2.2 系统功能设计 ······ 80
  3.2.3 系统数据库设计 ······ 81
  3.2.4 系统文件说明 ······ 87
 3.3 项目前端设计 ······ 88
  3.3.1 登录 ······ 88
  3.3.2 二手房房源管理 ······ 91
  3.3.3 楼盘信息管理 ······ 92
  3.3.4 房源信息管理 ······ 95
  3.3.5 楼盘动态管理 ······ 96
  3.3.6 认购管理 ······ 97
  3.3.7 销售管理 ······ 98
  3.3.8 认筹管理 ······ 99
  3.3.9 楼盘收藏管理 ······ 100
  3.3.10 系统管理与系统设置 ······ 100
 3.4 项目后端实现 ······ 105
  3.4.1 通用类 ······ 105
  3.4.2 登录 ······ 109
  3.4.3 二手房房源管理 ······ 109
  3.4.4 楼盘信息管理 ······ 111
  3.4.5 房源信息管理 ······ 113
  3.4.6 文件操作 ······ 114
  3.4.7 其他功能管理 ······ 116
 3.5 项目总结 ······ 116

第 4 章　购物车管理系统 ······ 118

 4.1 项目技术选型 ······ 118
  4.1.1 Spring Security ······ 118
  4.1.2 H2 数据库 ······ 122
  4.1.3 Thymeleaf ······ 129
  4.1.4 框架搭建 ······ 135
 4.2 项目前期准备 ······ 140
  4.2.1 项目需求说明 ······ 140
  4.2.2 系统功能设计 ······ 140
  4.2.3 系统数据库设计 ······ 140
  4.2.4 系统文件说明 ······ 141

4.3 项目前端设计 ··· 142
4.3.1 登录 ··· 142
4.3.2 注册 ··· 144
4.3.3 商品展示页面 ··· 145
4.3.4 购物车页面 ·· 147
4.3.5 通用导航 ··· 148
4.3.6 通用分页 ··· 149
4.3.7 安全校验错误页面 ·· 149
4.4 项目后端实现 ··· 150
4.4.1 登录与登录认证 ·· 150
4.4.2 注册与参数验证 ·· 157
4.4.3 异常处理 ··· 160
4.4.4 安全校验 ··· 163
4.4.5 商城首页 ··· 164
4.4.6 购物车与订单相关 ·· 165
4.5 项目总结 ·· 169

第 5 章 用户权限管理系统 ·· 171
5.1 项目技术选型 ··· 171
5.1.1 Shrio 权限认证框架 ·· 171
5.1.2 PostgreSQL 数据库 ··· 179
5.1.3 框架搭建 ··· 195
5.2 项目前期准备 ··· 198
5.2.1 项目需求说明 ··· 198
5.2.2 系统功能设计 ··· 198
5.2.3 系统数据库设计 ·· 199
5.2.4 系统文件说明 ··· 201
5.3 项目前端设计 ··· 202
5.3.1 登录 ··· 202
5.3.2 控制台首页 ·· 205
5.3.3 操作账号管理 ··· 207
5.3.4 菜单管理 ··· 208
5.3.5 按键管理 ··· 209
5.3.6 组织管理 ··· 209
5.3.7 角色管理 ··· 210
5.3.8 类型管理 ··· 210
5.3.9 分页展示 ··· 211
5.4 项目后端实现 ··· 211
5.4.1 登录认证和权限认证 ··· 211
5.4.2 验证码生成 ·· 215
5.4.3 操作账号管理 ··· 216
5.4.4 菜单管理 ··· 218
5.4.5 组织管理 ··· 219
5.4.6 其他管理 ··· 221
5.5 项目总结 ·· 221

第 6 章 使用小程序上报用户信息 ... 223

6.1 项目技术选型 ... 223
6.1.1 MyBatis-Plus 框架 ... 223
6.1.2 JWT ... 235
6.1.3 HbuilderX 简介 ... 237
6.1.4 小程序客户端项目搭建 ... 239
6.1.5 微信开发者工具 ... 241
6.1.6 后台服务框架搭建 ... 242

6.2 项目前期准备 ... 252
6.2.1 项目需求说明 ... 252
6.2.2 系统功能设计 ... 252
6.2.3 系统数据库设计 ... 253
6.2.4 系统文件说明 ... 258

6.3 项目前端设计 ... 259
6.3.1 首页 ... 259
6.3.2 我的 ... 260
6.3.3 微信一键登录 ... 260
6.3.4 完善信息 ... 261
6.3.5 底部导航栏 ... 262

6.4 项目后端实现 ... 263
6.4.1 JWT 登录认证 ... 263
6.4.2 登录与注册 ... 264
6.4.3 获取信息 ... 266
6.4.4 完善或修改信息 ... 267

6.5 项目总结 ... 268

第 7 章 模拟聊天室 ... 269

7.1 项目技术选型 ... 269
7.1.1 WebSocket ... 269
7.1.2 框架搭建 ... 280

7.2 项目前期准备 ... 282
7.2.1 项目需求说明 ... 282
7.2.2 系统功能设计 ... 282
7.2.3 系统数据库设计 ... 282
7.2.4 系统文件说明 ... 283

7.3 项目前端设计 ... 284
7.3.1 登录与退出 ... 284
7.3.2 聊天室主页面 ... 284
7.3.3 群发消息 ... 286
7.3.4 给指定用户单独发送消息 ... 286
7.3.5 上线与下线 ... 288

7.4 项目后端实现 ... 289
7.4.1 上线与下线 ... 289
7.4.2 发送消息 ... 292
7.4.3 获取当前在线用户列表 ... 293

7.5 项目总结 ... 294

第 1 章

Spring Boot入门

Spring发展至今已有十多年的历史。2001年10月，Rod Johnson 创作的 *Expert One-on-One J2EE* 将Spring的前身带入了开发者的视野，此书颠覆了当时开发者的认知。2003年，Spring正式诞生。此后，Spring一路突飞猛进，目前已成为Java开发中最流行、市场占比最高的框架。但是随着Spring的不断扩大，其弊端也逐渐显现，开发者面临越来越多的复杂配置和依赖管理，这使得Spring Boot应运而生。在本章中，我们首先对Spring Boot框架的历史、框架组成、特性等作一些简单的了解。

本章主要涉及的知识点有：

- Spring Boot的特性。
- Spring Boot的组成及特性。
- 配置开发环境。
- 使用Spring Boot快速创建项目。

1.1 Spring Boot介绍

Spring是轻量级的、灵活的企业级Java开发框架。Spring通过依赖注入和面向切面编程使得开发工作变得更高效简单。Spring的设计理念是非侵入性的，这使得Spring与业务代码实现了解耦。开发者在使用过程中几乎感受不到Spring框架的存在。随着Spring集成了越来越多的优秀框架，配置也变得越来越复杂，依赖关系也变得难以管理。Spring Boot的诞生在很大程度上解决了这些问题。

1.1.1 Spring Boot简介

在Spring官网上的Banner中滚动显示着这样几句话：

Spring makes Java simple.
Spring makes Java modern.
Spring makes Java productive.
Spring makes Java reactive.
Spring makes Java cloud-ready.

意思是Spring让Java开发更简单、创新、高效、灵活，为云服务做好准备。无疑，Spring使得开发效率得到了飞速提升。

在介绍Spring Boot之前，让我们先回顾一下目前Java Web开发框架的概况。Java Web的开发框架众多，有Tomcat、Jetty、 SpringStruts、Hibernate、MyBatis、JPA、JSP、Velocity、FreeMarker、Thymeleaf、Redis等；可选数据库也比较多，有MySQL、Oracle、MongoDB等；另外还有很多优秀的框架，虽然这些框架都是开源的，但将它们从零开始集成到一起，创建一个可以用来进行业务开发的框架仍然是一个非常浩大的工程。繁杂的配置管理、错综交叉的依赖关系、各框架之间版本兼容问题等，这些都让开发者头痛不已。而Spring Boot是解决这一问题的优选方案。它像一个容器一样，将所需要的组件集成到一起，为每个组件自动选择合适的版本，同时也会去除一些冲突的依赖，开发者不必再为这些问题苦恼。这也正像Spring宣扬的那样——Spring不重复造轮子，而是众多优秀框架的集成者。

Spring Boot是由Pivotal团队提供的全新框架，伴随着Spring 4.0出现，其主要目的就是简化框架搭建和开发过程。Spring Boot遵循的原则是"约定大于配置"，这样可以使开发者从复杂冗余的配置文件中解脱出来，专注业务本身。

Spring Boot有以下几点特性：

（1）能够创建独立的Spring应用

使用时只需要创建一个Application类，声明main方法，加上@SpringBootApplication即可。通过运行该类就可以启动一个Spring应用。

（2）内置了Tomcat、Jetty、Undertow等Web应用服务器，不需要再发布war包

Spring Boot已经内置了Web应用服务器，搭建时根据需要引入对应的依赖即可，不再需要打包过程。运行Application类时，Spring Boot会根据框架自动选择Web服务器，自动打包并发布到服务器。

（3）提供固定的starter依赖，简化框架搭建过程

starter依赖包含很多框架，使用starter引入依赖可免去版本冲突和选择问题，Spring Boot会自动选择合适版本，并避免版本冲突。

（4）自动配置所需的各项组件、框架

Spring Boot会自动引入Java Web应用中很多必需的依赖，并进行默认的配置。如果使用时不进行配置的话，则使用默认配置。

（5）提供量化、健康监测以及外部化配置

可以通过组件配置实现系统性能监控，对系统运行状况进行实时监测。

（6）不再需要任何XML配置

使用Spring Boot不再需要XML配置文件，可以通过properties文件配置，也可以通过yml文件配置。

1.1.2　Spring Boot核心模块

Spring Boot作为一个框架由众多模块组成，主要包括Spring Boot、Spring Boot AutoConfigure、Spring Boot Actuator、Spring Boot CLI、Spring Boot Devtools、Spring Boot Starters、Spring Boot Test等。下面逐一介绍这些模块的主要作用。

1. Spring Boot

Spring Boot的主模块，是其他模块的支撑，提供整个应用的启动类，内嵌Tomcat、Jetty、Undertow等Web应用服务器，负责加载配置信息，创建和刷新Spring容器的上下文内容。

2. Spring Boot AutoConfigure

Spring Boot AutoConfigure是自动配置功能。通过这个功能，Spring Boot会尝试为框架配置所需的参数。例如，在应用的classpath中存在HSQLDB依赖，但没有手动配置任何数据库连接，Spring Boot自动配置一个内存数据库。要开启该功能，在启动类上添加@EnableAutoConfiguration或@SpringBootApplication注解即可。

注意：以上两个注解只需注入其中一个即可。

这种自动配置是非侵入性的。开发者可以随时进行自定义配置。如果配置了自己的数据库连接，那么内置的数据库就会被覆盖。如果不希望Spring Boot进行某个组件的自动配置，可以只注解参数配置，将该组件排除在外。例如，如果不需要配置数据源，就可以使用exclude进行如下配置：

```
import org.springframework.boot.autoconfigure.SpringBootApplication;
import org.springframework.boot.autoconfigure.jdbc.DataSourceAutoConfiguration;
@SpringBootApplication(exclude = { DataSourceAutoConfiguration.class })
public class MyApplication {
}
```

如果要排除的组件不在classpath中，可以使用excludeName。@EnableAutoConfiguration和@SpringBootApplication都同时支持exclude和excludeName。

3. Spring Boot Actuator

Spring Boot Actuator是一个可以帮助监控系统数据的框架，包括应用程序的基本信息和健康监控、配置信息、请求跟踪等，支持通过HTTP端点或JMX两种形式进行配置。

Spring Boot Actuator模块提供了生产就绪的所有特性。官方推荐的引入该模块的方式是使用Starter依赖spring-boot-starter-actuator。

如果使用Maven，需要添加如下依赖：

```xml
<dependency>
    <groupId>org.springframework.boot</groupId>
    <artifactId>spring-boot-starter-actuator</artifactId>
</dependency>
```

如果使用Gradle，使用如下格式：

```
dependencies { compile("org.springframework.boot: spring-boot-starter-actuator") }
```

Spring Boot Actuator模块包含的端点如表1.1所示。

表1.1　Spring Boot Actuator端点列表

端点id	端点功能描述
auditevents	展示当前应用中的一些审核时间信息。需要配置AuditEventRepository类
beans	展示应用中的所有Bean对象
caches	展示可用缓存
conditions	显示配置类设置的条件、匹配与否及其原因
configprops	显示所有使用@ConfigurationProperties注解的对象
env	显示Spring环境参数
flyway	数据库版本控制工具
health	显示应用的健康信息
httptrace	显示HTTP追踪信息，需要配置HttpTraceRepository类
info	显示任意应用信息
integrationgraph	显示集成图解，需要引入spring-integration-core依赖
loggers	显示和配置应用的日志信息
liquibase	数据库版本控制工具
metrics	当前应用的指标信息
mappings	显示@RequestMapping注解对应的路径
quartz	Quartz定时任务列表
scheduledtasks	应用中的任务列表
sessions	查询和删除基于Spring session的用户session
shutdown	关闭应用。默认不启用
startup	显示启动信息，需要配置BufferingApplicationStartup
threaddump	执行线程转储

Web应用，例如Spring MVC、Spring WebFlux、Jersey这几种类型的应用，除了支持以上端点外，还支持如表1.2所示的几个端点。

表1.2　Web应用中Spring Boot Actuator支持的端点列表

端点id	端点功能描述
heapdump	返回HPROF格式的转储文件，需要HotSpot JVM
jolokia	显示JMX类，需要jolokia-core依赖
logfile	返回日志文件
prometheus	能够被Prometheus服务器抓取的信息，需要micrometer-registry-prometheus依赖

默认情况下，除shutdown端点以外的其他端点都被启用。可以在配置文件中对各端点的开启和关闭情况进行配置。例如，开启shutdown端点，在yml中的配置如下：

```yaml
management:
  endpoint:
    shutdown:
      enabled: true
```

如果只需要其中某个或某几个端点，可以禁用默认开启功能，然后按需配置端点。例如，只需要info端点，在yml中可以进行如下配置：

```yaml
management:
  endpoints:
    enabled-by-default: false
  endpoint:
    info:
      enabled: true
```

注意：禁用端点是将该端点从应用中彻底移除。如果只想更改端点是否对外暴露，则可以使用include和exclude属性。其中，include代表暴露端点，exclude代表不暴露端点。

默认情况下，Web应用只对外暴露health端点，可在yml中进行如下配置：

```yaml
management:
  endpoints:
    web:
      exposure:
        include: "*"
        exclude: "env,beans"
```

include的值为*号，代表暴露所有端点。

如果应用公开，则Spring Boot官方建议我们配置安全机制，可以使用Spring Security保证端点安全。引入Security依赖的代码如下：

```xml
<dependency>
    <groupId>org.springframework.boot</groupId>
    <artifactId>spring-boot-starter-security</artifactId>
</dependency>
```

在代码中进行如下配置：

```java
import org.springframework.boot.actuate.autoconfigure.security.servlet.EndpointRequest;
import org.springframework.context.annotation.Bean;
import org.springframework.context.annotation.Configuration;
import org.springframework.security.config.annotation.web.builders.HttpSecurity;
import org.springframework.security.web.SecurityFilterChain;

@Configuration(proxyBeanMethods = false)
public class MySecurityConfiguration {

    @Bean
    public SecurityFilterChain securityFilterChain(HttpSecurity http) throws Exception {
```

```
            http.requestMatcher(EndpointRequest.toAnyEndpoint())
                .authorizeRequests((requests) -> requests.anyRequest().permitAll());
            return http.build();
        }
    }
```

Actuator框架的默认访问路径是http://localhost:8080/actuator。如果想要修改访问路径,可以在配置文件中做如下修改:

```
#调整端点的前缀路径为/ management: endpoints: web: base-path: /
```

默认情况下端点不支持跨域,若想支持跨域,可通过如下配置修改:

```yaml
management:
  endpoints:
    web:
      cors:
        allowed-origins: "https://example.com"
        allowed-methods: "GET,POST"
```

4. Spring Boot Starters

Starters是一系列依赖的集合,使用Starters可以一站"购齐"所有所需的组件。例如,如果需要Spring以及Spring Data JPA组件,直接引入spring-boot-starter-data-jpa依赖即可。

这些组件的名称类似,命名结构为spring-boot-starter-*,其中*是具体应用的名称。这种命名结构可以使我们快速找到所需的组件。目前主流的开发工具,例如IntelliJ IDEA、Eclipse等,都支持使用对应快捷键进行提示。常用的一些Starter依赖如表1.3所示。

表1.3 Spring Boot Starter列表

Starter名称	主要功能描述
spring-boot-starter	核心Starter部件,包含自动配置支持、日志和yml配置
spring-boot-starter-activemq	使用Apache ActiveMQ组件实现的JMS消息
spring-boot-starter-amqp	包含Spring AMQP和Rabbit MQ
spring-boot-starter-aop	包含使用Spring AOP和AspectJ的面向切面编程
spring-boot-starter-artemis	使用Apache Artemis组件实现的JMS消息
spring-boot-starter-batch	包含Spring Batch
spring-boot-starter-cache	包含Spring框架的缓存
spring-boot-starter-data-cassandra	开源分布式数据库Cassandra
spring-boot-starter-data-cassandra-reactive	响应式的开源分布式数据库Cassandra
spring-boot-starter-data-couchbase	基于文档的数据库系统Couchbase
spring-boot-starter-data-couchbase-reactive	响应式的基于文档的数据库系统Couchbase
spring-boot-starter-data-elasticsearch	使用Elasticsearch搜索和分析引擎
spring-boot-starter-data-jdbc	使用Spring Data JDBC组件
spring-boot-starter-data-jpa	包含Spring Data JPA和Hibernate组件
spring-boot-starter-data-ldap	包含Spring Data LDAP组件

（续表）

Starter名称	主要功能描述
spring-boot-starter-data-mongodb	使用基于文档的数据库系统MongoDB
spring-boot-starter-data-mongodb-reactive	使用响应式的基于文档的数据库系统MongoDB
spring-boot-starter-data-neo4j	包含Neo4j图形数据库
spring-boot-starter-data-r2dbc	包含Spring Data R2DBC
spring-boot-starter-data-redis	包含使用键值对存储的Redis、Spring Data Redis组件和Lettuce客户端
spring-boot-starter-data-redis-reactive	包含使用键值对存储的Redis、响应式的Spring Data Redis组件和Lettuce客户端
spring-boot-starter-data-rest	使用Spring Data REST展示Spring Data数据存储库
spring-boot-starter-freemarker	使用FreeMarker视图建立MVC Web应用
spring-boot-starter-groovy-templates	使用Groovy模板视图建立MVC Web应用
spring-boot-starter-hateoas	使用Spring MVC和Spring HATEOAS建立基于超媒体的RESTful风格的Web应用
spring-boot-starter-integration	使用Spring Integration
spring-boot-starter-jdbc	使用基于HikariCP连接池的JDBC
spring-boot-starter-jersey	使用JAX-RS和Jersey建立RESTful风格的Web应用，也可以使用spring-boot-starter-web
spring-boot-starter-jooq	使用jOOQ与数据库建立连接，也可以使用spring-boot-starter-data-jpa或者spring-boot-starter-jdbc
spring-boot-starter-json	提供JSON支持
spring-boot-starter-jta-atomikos	使用Atomikos实现JTA事务
spring-boot-starter-mail	使用Java Mail和Spring框架中的email发送支持
spring-boot-starter-mustache	使用Mustache视图建立Web应用
spring-boot-starter-oauth2-client	使用Spring Security中的OAuth2/OpenID客户端连接
spring-boot-starter-oauth2-resource-server	使用Spring Security中的OAuth2/OpenID服务端连接
spring-boot-starter-quartz	使用Quartz实现任务
spring-boot-starter-rsocket	建立RSocke客户端和服务端
spring-boot-starter-security	使用Spring Security
spring-boot-starter-test	Spring Boot应用中的测试组件，包含JUnit Jupiter、Hamcrest和Mockito
spring-boot-starter-thymeleaf	使用Thymeleaf 视图建立MVC的Web应用
spring-boot-starter-validation	使用Hibernate Validator实现Java Bean Validation
spring-boot-starter-web	建立RESTful风格的Web应用，将内置的Tomcat作为默认的Web服务器
spring-boot-starter-web-services	使用Spring Web Services
spring-boot-starter-webflux	使用Spring框架的响应式Web支持建立WebFlux应用
spring-boot-starter-websocket	使用Spring框架中的WebSocket支持建立WebSocket应用

如果不需要某些组件，想要将它排除在外或替换成其他组件，也可以使用Starter名称，组件名称如表1.4所示。

表1.4 可排除或替换的Spring Boot Starter列表

Starter名称	功能描述
spring-boot-starter-jetty	使用Jetty作为servlet容器，也可以使用 spring-boot-starter-tomcat
spring-boot-starter-log4j2	使用Log4j2实现日志管理，也可以使用spring-boot-starter-logging
spring-boot-starter-logging	使用Logback实现日志管理，默认的日志管理
spring-boot-starter-reactor-netty	使用Reactor Netty作为内置的响应式HTTP服务器
spring-boot-starter-tomcat	使用Tomcat作为Servlet容器，这是 spring-boot-starter-web默认的Servlet容器
spring-boot-starter-undertow	使用undertow作为Servlet容器，也可以使用spring-boot-starter-tomcat

例如，不想用Tomcat而想用Jetty，可以在pom中进行如下配置：

```xml
<properties>
    <servlet-api.version>3.1.0</servlet-api.version>
</properties>
<dependency>
    <groupId>org.springframework.boot</groupId>
    <artifactId>spring-boot-starter-web</artifactId>
    <exclusions>
        <!-- Exclude the Tomcat dependency -->
        <exclusion>
            <groupId>org.springframework.boot</groupId>
            <artifactId>spring-boot-starter-tomcat</artifactId>
        </exclusion>
    </exclusions>
</dependency>
<!-- Use Jetty instead -->
<dependency>
    <groupId>org.springframework.boot</groupId>
    <artifactId>spring-boot-starter-jetty</artifactId>
</dependency>
```

5. Spring Boot CLI

Spring Boot CLI是一个命令行组件，使用它能够快速开启Spring应用。它能够运行Groovy脚本语言，还可以使用帮助命令提示用法。

```
$ spring help run
spring run - Run a spring groovy script

usage: spring run [options] <files> [--] [args]

Option                      Description
------                      -----------
--autoconfigure [Boolean]   Add autoconfigure compiler
                            transformations (default: true)
--classpath, -cp            Additional classpath entries
--no-guess-dependencies     Do not attempt to guess dependencies
--no-guess-imports          Do not attempt to guess imports
-q, --quiet                 Quiet logging
```

```
-v, --verbose              Verbose logging of dependency
                           resolution
--watch                    Watch the specified file for changes
```

如果使用Groovy编写了对应代码，就可以使用如下命令运行：

```
$ spring run hello.groovy -- --server.port=9000
```

端口是可选参数。

关于Spring Boot CLI的更多详细信息，可查看Spring Boot官方文档。

6. Spring Boot Devtools

Spring Boot Devtools是开发者工具，能够提升一些开发体验，例如日志打印、热部署、页面自动刷新等。

Spring Boot Devtools不仅支持本地应用，同时也支持远程应用。如果远程服务器环境安全，可以通过如下配置启用：

```xml
<build>
    <plugins>
        <plugin>
            <groupId>org.springframework.boot</groupId>
            <artifactId>spring-boot-maven-plugin</artifactId>
            <configuration>
                <excludeDevtools>false</excludeDevtools>
            </configuration>
        </plugin>
    </plugins>
</build>
```

注意：生产环境不建议开启该组件。

7. Spring Boot Test

Spring Boot提供了一系列的注解和依赖实现测试。实现测试功能需要引入spring-boot-test和spring-boot-test-autoconfigure依赖。大多数开发者直接使用Spring Boot提供的spring-boot-starter-test自动导入JUnit Jupiter、AssertJ、Hamcrest和一些其他相关的测试依赖库。

Spring应用需要提供Spring上下文环境，测试时也不例外，而创建Spring上下文相当烦琐，为此Spring Boot提供了@SpringBootTest注解。通过该注解可以直接对方法进行调用，而不需要再手动创建Spring Context环境。示例代码如下：

```java
import org.junit.jupiter.api.Test;

import org.springframework.boot.test.context.SpringBootTest;
import org.springframework.context.annotation.Import;

@SpringBootTest
class MyTests {
    @Test
    void exampleTest() {
        // ...
    }
}
```

如果需要为测试环境单独创建配置，可以使用@TestConfiguration创建配置类，然后使用@Import注解导入配置类。示例代码如下：

```
@SpringBootTest
@Import(MyTestsConfiguration.class)
class MyTests {

}
@TestConfiguration
class MyTestsConfiguration{

}
```

Spring Boot提供专门应用于JSON、Spring MVC、Spring WebFlux、Data JPA、MongoDB、Redis等的测试环境配置信息，通过对应注解即可引入。更多详细信息读者可查看Spring Boot官方文档。

1.2 开发环境配置

工欲善其事，必先利其器。本节主要讲解常用工具的安装与配置，包括JDK开发库、IntelliJ IDEA开发工具、Maven构建工具以及MySQL数据库。

1.2.1 安装与配置JDK

JDK支持多个操作系统平台，对应的官方下载地址为https://www.oracle.com/java/technologies/downloads/#java17。这里以Windows 10 64位操作系统为例，演示JDK17版本的安装与配置。

1. 安装JDK

步骤01 打开JDK17的下载页面，找到对应的版本进行下载，如图1.1所示。

Linux　macOS　**Windows**

Product/file description	File size	Download
x64 Compressed Archive	170.64 MB	https://download.oracle.com/java/17/latest/jdk-17_windows-x64_bin.zip (sha256)
x64 Installer	151.99 MB	https://download.oracle.com/java/17/latest/jdk-17_windows-x64_bin.exe (sha256)
x64 MSI Installer	150.88 MB	https://download.oracle.com/java/17/latest/jdk-17_windows-x64_bin.msi (sha256)

图1.1　JDK下载页面

步骤02 下载完成后，打开文件进行安装。在对话框中直接单击“下一步”按钮，如图1.2所示。

步骤03 选择好安装路径，然后单击“下一步”按钮，如图1.3所示。

步骤04 提示安装完成后，单击“关闭”按钮，如图1.4所示。

图1.2　安装JDK（1）

图1.3　安装JDK（2）

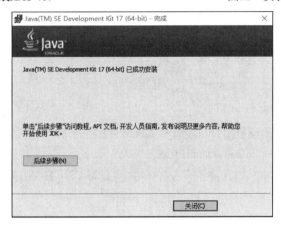

图1.4　安装JDK（3）

步骤 05　JDK9及以后的版本不再附带jre的安装，如果需要jre，可以进行手动配置。首先按快捷键Windows+R打开cmd，切换到JDK安装目录。然后输入如下命令，执行完毕后没有任何输出，表示执行成功，如图1.5所示。

图1.5　生成jre

步骤 06　打开JDK17的安装目录，可以看到生成了jre目录，如图1.6所示。

2. 配置JDK

步骤 01　进入Windows设置页面，输入"高级系统"，选择"查看高级系统设置"选项，如图1.7所示。

步骤 02　配置环境变量。

首先新建JAVA_HOME变量，变量值为jdk安装目录，如图1.8所示。

图1.6　jre目录

图1.7　Windows设置页面

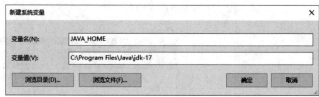

图1.8　配置环境变量（1）

其次打开Path变量设置窗口（没有看到Path的话新建一个），加入**%JAVA_HOME%\bin**和**%JAVA_HOME%\jre\bin**，如图1.9所示。

然后新建CLASSPATH变量，变量值为.;%JAVA_HOME%\lib;%JAVA_HOME%\lib\tools.jar（注意最前面的"."），如图1.10所示。

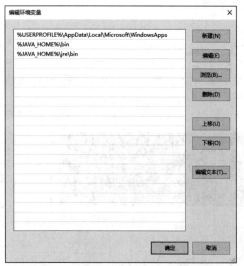

图1.9　配置环境变量（2）

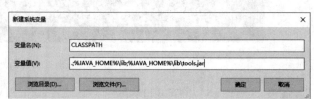

图1.10　配置环境变量（3）

步骤03 进行测试。打开cmd窗口，输入java -version命令，如果显示正确的JDK信息，表示安装和配置成功，如图1.11所示。

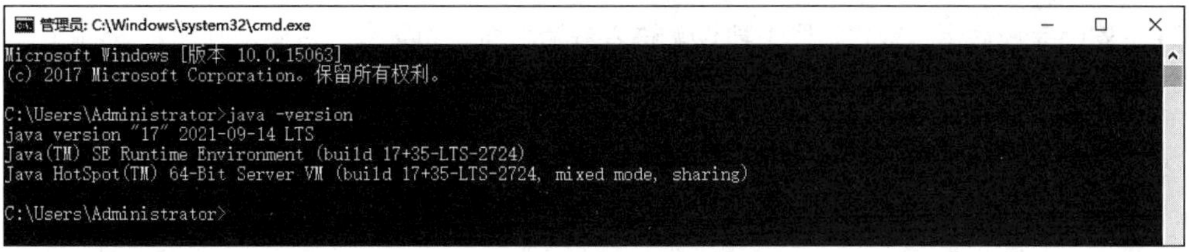

图1.11　测试JDK

1.2.2　安装IntelliJ IDEA

IntelliJ IDEA是目前比较受欢迎的开发工具，它集成了众多插件，环境配置也相对简单，容易上手，支持各种类型的语言和开发框架。接下来，我们演示如何在Windows 10中安装该开发工具。

步骤01 首先到IntelliJ IDEA官网下载最新版本，目前最新版本为2021.2.2，官方下载地址为：https://www.jetbrains.com/idea/download/，下载页面如图1.12所示。

图1.12　下载IDEA（1）

步骤02 下载完成后，双击安装包进行安装，如图1.13所示。

步骤03 每一步都选择默认选项，直接单击"Next"按钮，进入安装进程，如图1.14所示。

图1.13　安装IDEA（2）

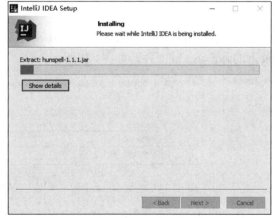

图1.14　安装IDEA（3）

步骤 04 安装完成后运行软件，同意协议，进入激活页面。如果有激活码，则输入激活码；如果没有激活码，该软件支持30天试用，可跳过该步骤，随后再激活。激活页面如图1.15所示。

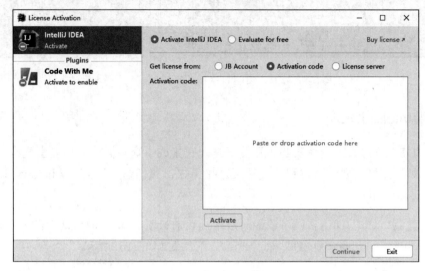

图1.15 激活IDEA

1.2.3 安装Maven

目前，Maven的最新版本是3.8.2。官方下载地址为https://maven.apache.org/download.cgi。下载页面如图1.16所示，选择ZIP格式的压缩包进行下载。

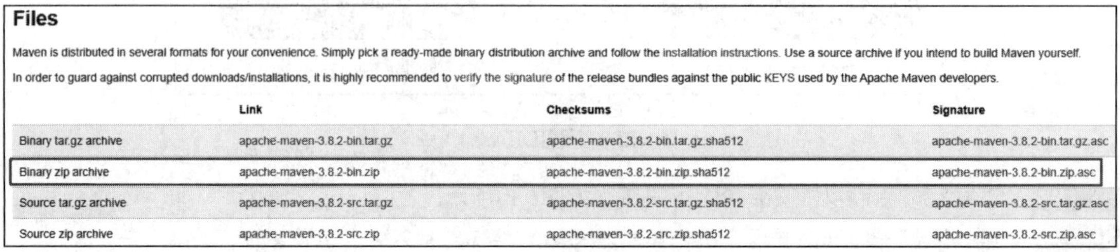

图1.16 Maven下载页面

Maven目前是免安装版，下载完成后直接解压到指定路径即可。解压完成后，打开conf目录下的settings.xml文件，如图1.17所示。

图1.17 Maven安装目录

在settings.xml文件中加入如下内容，配置本地仓库地址。

```
<localRepository>E:\repository</localRepository>
```

Maven默认使用自带的远程仓库地址，如果有自己在本地搭建的Maven仓库，可以将远程仓库地址配置成自己的地址。

```
<mirror>
    <id>nexus-sifu</id>
    <name>internal nexus repository</name>
    <!--镜像采用配置好的组的地址-->
    <url>http://your IP or host and port/repository/alibaba-group/</url>
    <mirrorOf>*</mirrorOf>
</mirror>
```

以上操作完成后，需要在IntelliJ IDEA中配置Maven信息。打开IDEA，在菜单栏中单击File→Settings命令，搜索Maven，打开Maven配置页面，配置本地Maven信息，如图1.18所示。

配置完成后，该IDEA项目就可以使用本地的Maven仓库了。

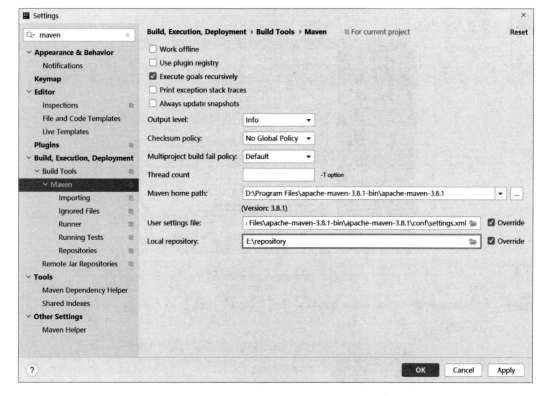

图1.18　IDEA配置Maven

1.2.4　安装MySQL

读者可以免费下载MySQL 8.0版本。本书使用的是MySQL 8.0.26版本。下载地址为https://dev.mysql.com/downloads/windows/installer/。

步骤01　进入MySQL官网的下载页面之后，在Select Operating System（选择操作系统）下拉列表框中选择Microsoft Windows，单击社区版对应的Download按钮，如图1.19所示。

下载好的安装文件如图1.20所示。

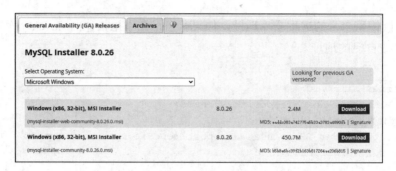

图1.19　MySQL下载页面

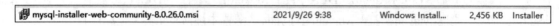

图1.20　MySQL 8.0.26安装文件

步骤02 双击MySQL安装程序，进入License Agreement窗口，如图1.21所示。

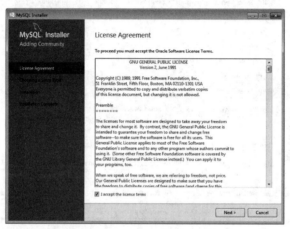

图1.21　License Agreement窗口

步骤03 勾选I accept the license terms复选框，单击"Next"按钮进入Choosing a Setup Type窗口，如图1.22所示。

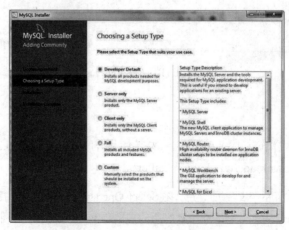

图1.22　Choosing a Setup Type窗口

步骤04 选中Developer Default单选框，单击"Next"按钮进入Check Requirements窗口，如图1.23所示。

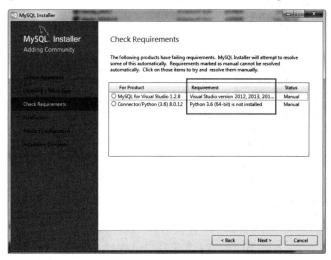

图1.23　Check Requirements窗口

步骤05 单击"Next"按钮，弹出提示对话框提示需要手动安装的组件，如图1.24所示。

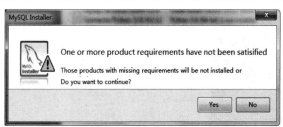

图1.24　Requirements提示

步骤06 手动安装组件后，单击"Next"按钮，进入Installation窗口，如图1.25所示。

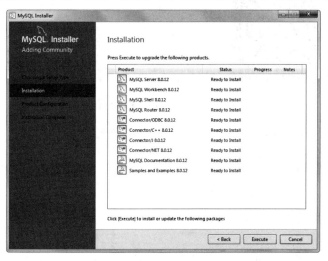

图1.25　Installation窗口

步骤07 单击"Execute"按钮，安装完成后的Installation窗口如图1.26所示。

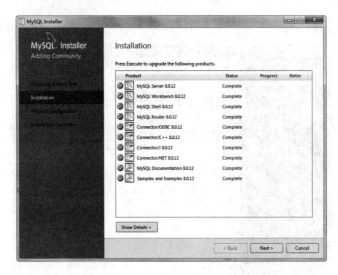

图1.26　安装完成后的Installation窗口

至此，MySQL 8.0安装完毕，接下来将介绍MySQL 8.0的配置。

前面的设置部分可以直接选择默认设置，这里主要介绍如何设置用户名和密码。Accounts and Roles窗口如图1.27所示。

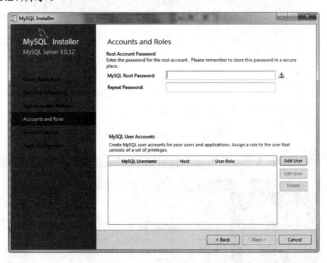

图1.27　Accounts and Roles窗口

步骤01 在MySQL Root Password和Repeat Passord中输入root账户的密码，单击Add User按钮，打开如图1.28所示对话框。

步骤02 输入用户名、主机、角色、密码等信息，单击OK按钮，就会成功添加一个账户，如图1.29所示。

步骤03 接下来全部选择默认选项，一直单击"Next"按钮，直到安装完成，如图1.30所示。

图1.28　MySQL User Details对话框

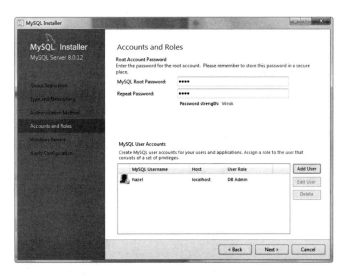

图1.29　成功添加一个账户

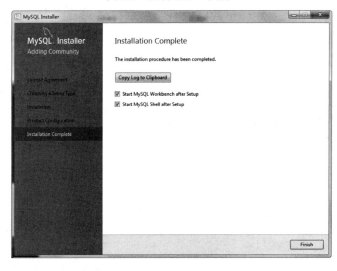

图1.30　安装完成

步骤04 最后配置Path变量。参考之前JDK环境变量的配置步骤，打开环境变量页面，找到Path变量，编辑并新增一条变量，变量值为C:\Program Files\MySQL\MySQL Server 8.0\bin。这样MySQL数据库的Path变量就添加好了，可以直接在DOS窗口输入"mysql"命令了。

1.3　使用Spring Boot

1.3.1　快速搭建一个项目

本节我们使用IntelliJ IDEA集成开发环境快速创建一个基于Maven的Spring Boot项目。

步骤01 首先打开IDEA，依次单击File→New→Project命令，进入New Project页面，如图1.31所示。

步骤 02　选中Spring Initializr，Project SDK设置为version 17，默认采用Spring官方的Spring Initializr服务，URL为https://start.spring.io/。

步骤 03　单击"Next"按钮进入Spring Initializr Project Settings设置页面，设置Type为Maven，Language为Java，Packaging为Jar，如图1.32所示。

图1.31　新建Spring Boot项目

图1.32　设置Spring Boot项目

步骤 04　单击"Next"按钮进入Spring Initializr初始化项目页面，选择项目组件，如图1.33所示。

步骤 05　单击"Next"按钮进入最后一步，在该页面输入项目名称，选择项目代码存放路径，单击"Finish"按钮，这样我们就快速得到了一个基于Maven的Spring Boot项目，如图1.34所示。

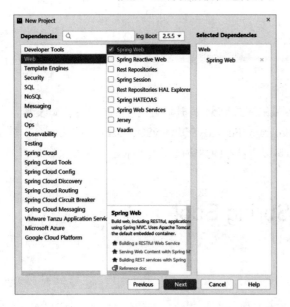
图1.33　选择项目组件

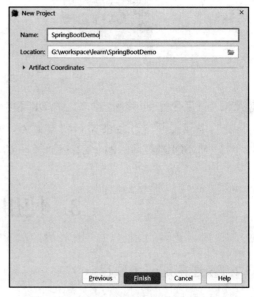
图1.34　设置项目名称和代码存放位置

1.3.2 运行发布项目

在IDEA中运行Spring Boot项目非常简单，且有多种方式。第一，通过底部的Services菜单；第二，通过右上角的运行设置；第三，直接运行Application入口类的main方法。

1. 通过底部的Services菜单

步骤01 依次单击View→Tool Windows→Services，如图1.35所示，打开IDEA的Services窗口，打开后可以在底部看到对应的窗口，如图1.36所示。

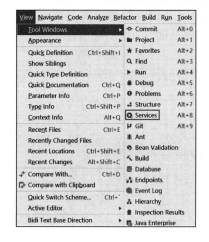

图1.35　打开Services窗口的菜单项

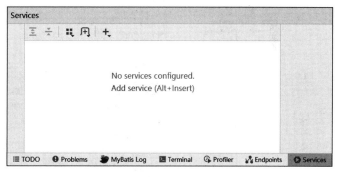

图1.36　Services窗口

步骤02 在窗口中单击Add service文字链接，选择Spring Boot项目，如图1.37所示。

选中Spring Boot后，IDEA会自动添加满足条件的Spring Boot启动类，如图1.38所示。

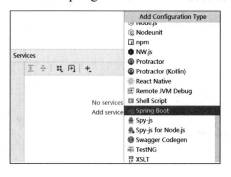

图1.37　添加Spring Boot Service

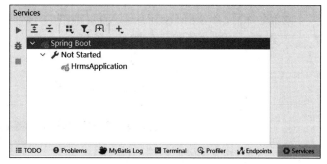

图1.38　添加Spring Boot启动类

接下来就可以通过左侧边栏上的运行、Debug模式运行等按钮运行项目了。

2. 通过右上角的运行设置

项目创建完成后，IDEA会自动生成对应的运行设置，并显示在页面右上角，运行设置的名字就是该项目的启动类名称，如图1.39所示。开发者可以使用对应的按钮进行运行、Debug模式运行等操作。

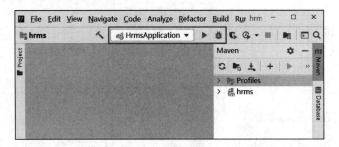

图1.39　运行设置

如果未生成该配置，也可以手动生成。单击上述页面右上角的下拉框，弹出选择项。选择Edit Configurations，在弹出框中选择Spring Boot，打开配置页面，如图1.40所示。

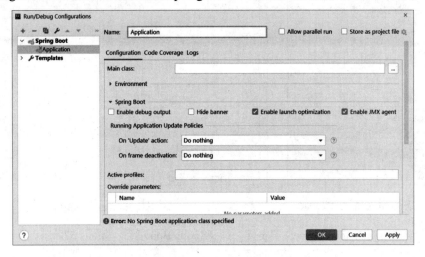

图1.40　运行设置配置页

输入名称后，单击Main Class输入框右侧的 按钮，在弹出的页面中选择Spring Boot Class，单击"OK"按钮确定选择，如图1.41所示。

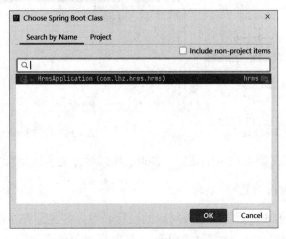

图1.41　选择Spring Boot Main Class

选中后，Main Class会显示在主设置页中，如图1.42所示。

第 1 章 Spring Boot入门 | 23

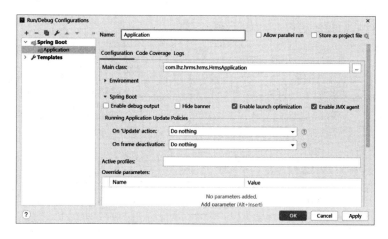

图1.42 运行设置

单击"OK"按钮，完成设置，可以看到右上角出现了我们刚刚设置的Application，如图1.43所示。

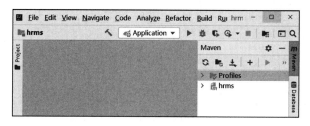

图1.43 运行设置

3. 通过Main方法

除了以上两种方式，也可以直接在启动类上通过鼠标右键选择运行按钮，如图1.44所示。

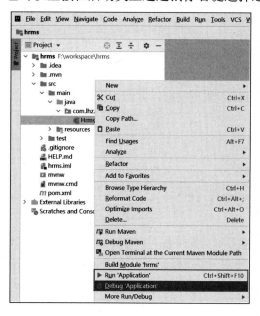

图1.44 通过鼠标右键运行项目

项目成功运行后，控制台会打印如下输出。

```
  .   ____          _            __ _ _
 /\\ / ___'_ __ _ _(_)_ __  __ _ \ \ \ \
( ( )\___ | '_ | '_| | '_ \/ _` | \ \ \ \
 \\/  ___)| |_)| | | | | || (_| |  ) ) ) )
  '  |____| .__|_| |_|_| |_\__, | / / / /
 =========|_|==============|___/=/_/_/_/
 :: Spring Boot ::       v2.5.5

   2021-02-03 10:33:25.224  INFO 17321 --- [           main] 
o.s.b.d.s.s.SpringAppplicationExample    : Starting SpringAppplicationExample using
Java 1.8.0_232 on mycomputer with PID 17321 (/apps/myjar.jar started by pwebb)
   2021-02-03 10:33:25.226  INFO 17900 --- [           main] 
o.s.b.d.s.s.SpringAppplicationExample    : No active profile set, falling back to
default profiles: default
   2021-02-03 10:33:26.046  INFO 17321 --- [           main] 
o.s.b.w.embedded.tomcat.TomcatWebServer  : Tomcat initialized with port(s): 8080
(http)
   2021-02-03 10:33:26.054  INFO 17900 --- [           main] 
o.apache.catalina.core.StandardService   : Starting service [Tomcat]
   2021-02-03 10:33:26.055  INFO 17900 --- [           main] 
org.apache.catalina.core.StandardEngine  : Starting Servlet engine: [Apache
Tomcat/9.0.41]
   2021-02-03 10:33:26.097  INFO 17900 --- [           main] 
o.a.c.c.C.[Tomcat].[localhost].[/]       : Initializing Spring embedded
WebApplicationContext
   2021-02-03 10:33:26.097  INFO 17900 --- [           main] 
w.s.c.ServletWebServerApplicationContext : Root WebApplicationContext:
initialization completed in 821 ms
   2021-02-03 10:33:26.144  INFO 17900 --- [           main] 
s.tomcat.SampleTomcatApplication         : ServletContext initialized
   2021-02-03 10:33:26.376  INFO 17900 --- [           main] 
o.s.b.w.embedded.tomcat.TomcatWebServer  : Tomcat started on port(s): 8080 (http) with
context path ''
   2021-02-03 10:33:26.384  INFO 17900 --- [           main] 
o.s.b.d.s.s.SpringAppplicationExample    : Started SampleTomcatApplication in 1.514
seconds (JVM running for 1.823)
```

如果运行失败，控制台打印如下输出，并给出错误提示。

```
***************************
APPLICATION FAILED TO START
***************************

Description:

Embedded servlet container failed to start. Port 8080 was already in use.

Action:

Identify and stop the process that's listening on port 8080 or configure this
application to listen on another port.
```

如果使用Spring Cloud微服务框架，这时Spring Boot项目比较多，选择第一种方式更直观、更方便。

使用Maven发布项目也非常简单，我们只需打开右侧Maven窗口，选择package或install选项，如图1.45所示。

执行打包命令后，就可以在项目目录中看到对应的target目录。该目录下生成了对应的JAR包或WAR，如图1.46所示。开发者将打包生成的文件传输到服务器上运行即可。

图1.45　Maven构建窗口

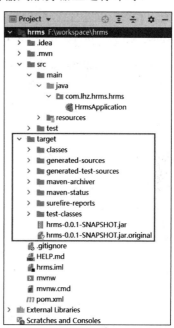

图1.46　target目录

1.4　Spring Boot源码分析

1.4.1　入口类@SpringBootApplication

在前面快速创建的入口类中，添加了@SpringBootApplication注解，这个注解的定义如下：

```
package org.springframework.boot.autoconfigure;
import ...
@Target({ElementType.TYPE})
@Retention(RetentionPolicy.RUNTIME)
@Documented
@Inherited
@SpringBootConfiguration
@EnableAutoConfiguration
@ComponentScan(
    excludeFilters = {@Filter(
    type = FilterType.CUSTOM,
    classes = {TypeExcludeFilter.class}
), @Filter(
```

```
        type = FilterType.CUSTOM,
        classes = {AutoConfigurationExcludeFilter.class}
)}
)
public @interface SpringBootApplication {...}
```

从上面的代码中可以看到，@SpringBootApplication这个注解封装了@SpringBootConfiguration、@EnableAutoConfiguration、@ComponentScan这3个注解。

- @EnableAutoConfiguration：自动配置注解，开启自动配置机制。
- @ComponentScan：设置扫描类包报名。
- @SpringBootConfiguration：注册其他的组件或导入外部的配置类，也可以使用@Configuration注解。

1. @EnableAutoConfiguration

自动配置注解@EnableAutoConfiguration的源码如下：

```
package org.springframework.boot.autoconfigure;
import ...
@Target({ElementType.TYPE})
@Retention(RetentionPolicy.RUNTIME)
@Documented
@Inherited
@AutoConfigurationPackage
@Import({AutoConfigurationImportSelector.class})
public @interface EnableAutoConfiguration {...}
```

@Import导入由@Configuration标注的配置类。如果存在多个配置类，该注解会自动将这些配置类综合成一个配置结果。

@EnableAutoConfiguration注解使得Spring Boot更智能化。Spring Boot会根据添加的依赖组件尝试猜测所需的配置。

2. @ComponentScan

该注解的源码如下：

```
package org.springframework.context.annotation;
import ...
@Retention(RetentionPolicy.RUNTIME)
@Target({ElementType.TYPE})
@Documented
@Repeatable(ComponentScans.class)
public @interface ComponentScan {...}
```

通过该注解设置Spring要自动扫描的包路径。如果不设置的话，会自动扫描当前启动类所在的包路径。

3. @SpringBootConfiguration

该注解的源码如下：

```
package org.springframework.boot;
import ...
@Target({ElementType.TYPE})
@Retention(RetentionPolicy.RUNTIME)
@Documented
@Configuration
@Indexed
public @interface SpringBootConfiguration {...}
```

可以看到@SpringBootConfiguration注解封装了@Configuration注解。而@Configuration注解又封装了@Component注解：

```
package org.springframework.context.annotation;
import ...
@Target({ElementType.TYPE})
@Retention(RetentionPolicy.RUNTIME)
@Documented
@Component
public @interface Configuration {...}
```

@Configuration注解标识了该类是个配置类。使用这种方式代替配置文件，极大地简化了配置文件，提高了可阅读性。

1.4.2 深入理解自动配置

Spring Boot自动配置功能可以根据不同情况来决定Spring应该配置哪些配置。例如，Spring JPA的依赖包在类路径里，并且DataSource也存在，就自动配置JpaRepositoriesAutoConfiguration，开发者无须进行额外配置，直接定义对应的实体类和数据访问层即可开始实现业务逻辑。又如，Thymeleaf依赖包在类路径里，那么Spring会自动配置Thymeleaf的模板解析器、视图解析器、模板引擎等。这些自动配置的实现都是基于Spring提供的条件化配置功能，主要依赖的注解就是@Conditional。

首先我们看一下不需要条件直接注入的自定义配置。下面是使用Java Config方式声明的MyBatisPlusConfig的配置。

```
@Configuration
public class MyBatisPlusConfig {
    //分页用
    @Bean
    public PaginationInterceptor paginationInterceptor(){
        PaginationInterceptor paginationInterceptor = new PaginationInterceptor();
        return paginationInterceptor;
    }
    @Bean
    public EasySqlInjector easySqlInjector () {
        return new EasySqlInjector();
    }
}
```

该配置类中使用@Configuration注解声明这是一个配置类，在配置类中使用@Bean注入

PaginationInterceptor、EasySqlInjector对象。这种方式比XML文件简洁明了，更易阅读。

为了让这种方式更灵活，Spring Boot提供了更多注解来帮助实现自动配置，其中包括条件注解@Conditional。这个注解从Spring 4.0之后开始引入，可以作用在创建Bean的类或者方法上，作用是按照一定的条件进行判断，满足条件的话，给容器注册对应的Bean对象。

首先看一下@Conditional这个注解的源码：

```
@Target({ElementType.TYPE, ElementType.METHOD})
@Retention(RetentionPolicy.RUNTIME)
@Documented
public @interface Conditional {
    /**
     * All {@link Condition Conditions} that must {@linkplain Condition#matches match}
     * in order for the component to be registered.
     */
    Class<? extends Condition>[] value();
}
```

它所接收的参数是Condition对象数组。Condition是Spring提供的一个接口，用于判断条件是否满足。

```
@FunctionalInterface
public interface Condition {
    /**
     * Determine if the condition matches.
     * @param context the condition context
     * @param metadata metadata of the {@link
     org.springframework.core.type.AnnotationMetadata class}
     * or {@link org.springframework.core.type.MethodMetadata method} being checked
     * @return {@code true} if the condition matches and the component can be registered,
     * or {@code false} to veto the annotated component's registration
     */
    boolean matches(ConditionContext context, AnnotatedTypeMetadata metadata);
}
```

matches为比对方法，当比对结果为true时注入该Bean对象。开发者需要继承Condition接口编写自己的实现类，然后在@Conditional注解的value值中引入该实现类，才能实现条件注入。但实际应用中，这种方式操作起来非常烦琐，因此Spring Boot基于@Conditional注解进一步封装了@ConditionalOnXXX一系列的注解。对于这些注解，开发者可以不用再实现Condition接口，而是直接在@ConditionalOnXXX注解的value值中写入判断的条件（例如class名称、path路径、properties配置等），提高了开发效率。

常用到的@ConditionalOnXXX注解有以下几种：

- @ConditionalOnClass、@ConditionalOnMissingClass：根据指定类是否存在决定是否实例化对象。例如，当SomeService.class存在时，实例化SomeServiceConfiguration对象。

```
@Configuration(proxyBeanMethods = false)
// Some conditions ...
```

```
public class MyAutoConfiguration {
    // Auto-configured beans ...
    @Configuration(proxyBeanMethods = false)
    @ConditionalOnClass(SomeService.class)
    public static class SomeServiceConfiguration {

        @Bean
        @ConditionalOnMissingBean
        public SomeService someService() {
            return new SomeService();
        }
    }
}
```

- @ConditionalOnBean、ConditionalOnMissingBean：根据指定Bean对象是否存在决定是否实例化对象。例如，当容器中不存在SomeService对象时，才会实例化SomeService对象。

```
@Configuration(proxyBeanMethods = false)
public class MyAutoConfiguration {
    @Bean
    @ConditionalOnMissingBean
    public SomeService someService() {
        return new SomeService();
    }
}
```

- @ConditionalOnProperty：根据Spring环境配置属性决定是否实例化对象。可以使用prefix和name属性指定Spring配置项的名称。
- @ConditionalOnResource：资源存在时才会实例化Bean对象。资源名称可以使用Spring约定的命名方式，例如：file:/home/user/test.dat。
- @ConditionalOnExpression：根据Spring表达式的结果决定是否实例化对象。
- @ConditionalOnWebApplication、@ConditionalOnNotWebApplication：根据当前应用是否是Web应用决定是否实例化对象。

1.5 本章小结

本章主要介绍了Spring Boot的特性、核心模块内容、如何快速使用以及部分源码分析，并讲解了如何从零开始安装与配置一个常用的开发环境，包括JDK、Maven、IntelliJ IDEA以及MySQL的安装与配置。通过本章学习，读者应该能够非常熟练地搭建开发环境，快速开启一个Spring Boot项目。通过核心模块剖析和源码讲解，读者对Spring Boot的组成框架和实现原理也应有一定的了解。这些都将为后面的案例讲解打下基础。

第 2 章

员工管理系统

在第1章中，我们介绍了Spring Boot核心知识点以及常用软件的安装与配置。从这一章开始，我们将综合利用常见的工具和组件进行项目开发实战。本章主要介绍如何通过Spring Boot，利用Maven集成Spring MVC、MyBatis以及MySQL数据库建立后台的管理框架，并通过集成LayUI前端框架实现员工管理系统的功能，包括员工基本信息管理、部门管理、工资管理、考勤管理等。

本章主要涉及的知识点有：

- 如何使用Spring Boot集成MyBatis。
- 如何使用Spring Boot配置MySQL数据库。
- 如何使用MyBatis完成基本的增、删、改、查。
- 如何使用LayUI前端框架与后台接口进行数据交互。

2.1 项目技术选型

架构对于一个项目的开发起着至关重要的作用。选择合适的框架能让开发效率事半功倍。本章涉及的员工管理系统选用的框架是常见的SSM框架，即Spring、Spring MVC和MyBatis。

2.1.1 MyBatis

MyBatis源于Apache的开源项目iBatis，2010年改名为MyBatis。

MyBatis是一个非常优秀的持久层框架，它支持自定义SQL、存储过程以及高级映射。它小且简单，几乎消除了所有的JDBC代码、设置参数以及处理结果集的工作，基本做到JDBC无感。MyBatis可以通过简单的XML或注解来配置和映射原始类型、接口和Java POJO。MyBatis非常容易上手，并且还有一些配套的插件，例如MyBatis-Plus、MyBatis-Generator等，使得开发工作更加便捷。

MyBatis的大体工作流程如图2.1所示。

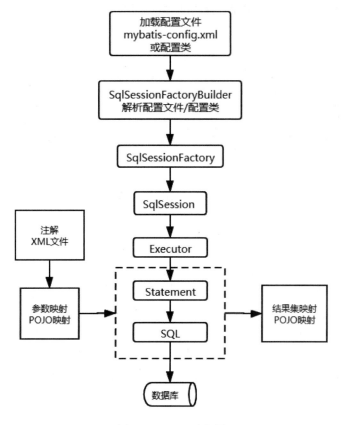

图2.1 MyBatis流程图

首先使用MyBatis自己提供的IO组件加载config配置文件，然后通过SqlSessionFactoryBuilder解析配置信息，得到SqlSessionFactory，再由SqlSessionFactory开启session得到SqlSession，接下来使用SqlSession调用对应的Mapper接口。这里MyBatis会自动代理，执行Executor部分，执行时，会根据解析的参数和Java POJO相关的映射信息，得到SQL语句；执行完毕后，得到对应的结果集。MyBatis同样会根据XML配置文件将结果集映射为对应的对象，常常是Java POJO。

上述步骤是传统开发方式的步骤。如果使用Spring Boot，那么MyBatis的config配置可以替换为配置类，由Spring Boot使用@Configuration注解进行扫描。

为了便于理解，可以看一下MyBatis的源码。MyBatis的源码可到GitHub上下载，地址是https://github.com/mybatis/mybatis-3。为了便于理解，可以将MyBatis的源码归类为几部分组成的框架，框架结构如图2.2所示。

API接口部分有MyBatis提供的可供调用的接口，包括数据库的增、删、改、查和配置信息的获取。这是直接面向使用者开放的接口。

数据处理部分包括参数解析、SQL解析、SQL执行和结果映射等部分。参数解析包括参数绑定和参数映射等；SQL解析包括SQL获取、动态SQL等；SQL执行部分会根据前边的参数解析和SQL解析生成对应的结果集，并根据配置文件映射到对应的对象中。

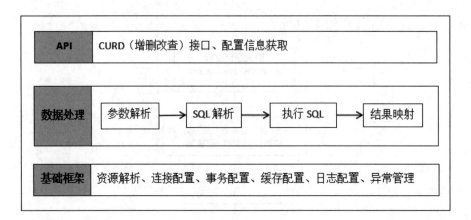

图2.2　MyBatis框架结构

基础框架部分包含IO资源加载与解析、数据库连接配置、事务配置、缓存配置、日志配置和异常管理。这里重点说一下MyBatis的日志配置。MyBatis本身只提供了日志工厂，如果想要实现日志功能需要借助第三方日志工具，目前使用较多的有Log4j、log4j2、SLF4J等，使用时只需引入对应的JAR包，做些简单的配置，即可与MyBatis无缝结合。

虽然MyBatis简单灵活，容易上手，扩展性强，有很多兼容的第三方插件，但是它依然存在一些问题：由于MyBatis基于SQL语句，所以SQL语句的工作量较大，对开发人员的SQL功底有较高的要求；由于MyBatis依赖SQL，所以数据库的移植性差，无法实现跨数据库。实际应用过程中，应根据业务需要、人力资源情况等综合选择对应的ORM（Object Relational Mapping，对象关系映射）框架。

2.1.2　框架搭建

如今的IDE开发工具功能强大且人性化，我们可以选择自行手写pom文件完成搭建，也可以利用开发工具提供的便捷途径来搭建。下面我们使用IntelliJ IDEA这个开发工具进行本章项目的开发环境搭建。

本章项目选取的是员工管理系统，以下为该项目的技术框架部分：

- 开发框架：Spring Boot。
- 数据库：MySQL。
- 后台框架：Spring、SpringMVC、MyBatis。
- 动态页面技术：JSP。
- 前端框架：LayUI。

接下来使用IntelliJ IDEA完成该框架的搭建。

1．新建项目

步骤01　打开IntelliJ IDEA，单击New Project按钮，选择新建一个项目，如图2.3所示。

步骤02　在弹出的New Project窗口中单击Spring Initializr选项，选择对应的JDK版本，在Choose starter service URL下选择Default，单击"Next"按钮，如图2.4所示。

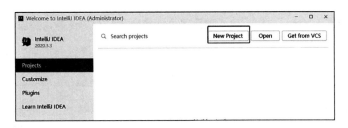

图2.3　IntelliJ IDEA首页

步骤 03　在Spring Initializr Project Settings窗口选择对应的构建信息，Type选择Maven，单击"Next"按钮，如图2.5所示。

图2.4　新建Spring Boot项目

图2.5　设置Spring Boot项目信息

步骤 04　在弹出的组件选择页面中，单击Web选项，勾选Spring Web复选框，如图2.6所示，这样就会自动加入Spring MVC相关的JAR包。

步骤 05　继续停留在该页面中，单击SQL选项，勾选MyBatis Framework和MySQL Driver复选框，单击"Next"按钮，如图2.7所示，这样MyBatis和MySQL的驱动就直接加载进来了。

图2.6　新建项目选择Web项

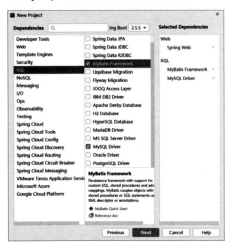

图2.7　选择SQL项

步骤 06 在弹出的保存信息页面中，输入项目名称，选择代码存放位置，然后单击"Finish"按钮完成创建，如图2.8所示。

图2.8 设置项目名称

新建完成后，打开项目。再打开对应的pom文件，可以看到对应的框架都已经引入进来了。

```xml
<parent>
    <groupId>org.springframework.boot</groupId>
    <artifactId>spring-boot-starter-parent</artifactId>
    <version>2.5.5</version>
    <relativePath/> <!-- lookup parent from repository -->
</parent>
<groupId>com.lhz</groupId>
<artifactId>hrms</artifactId>
<version>0.0.1-SNAPSHOT</version>
<name>hrms</name>
<description>hrms</description>
<properties>
    <java.version>17</java.version>
</properties>
<dependencies>
    <dependency>
        <groupId>org.springframework.boot</groupId>
        <artifactId>spring-boot-starter-web</artifactId>
    </dependency>
    <dependency>
        <groupId>org.mybatis.spring.boot</groupId>
        <artifactId>mybatis-spring-boot-starter</artifactId>
        <version>2.1.4</version>
    </dependency>

    <dependency>
        <groupId>mysql</groupId>
        <artifactId>mysql-connector-java</artifactId>
        <scope>runtime</scope>
    </dependency>
```

```xml
    <dependency>
        <groupId>org.springframework.boot</groupId>
        <artifactId>spring-boot-starter-test</artifactId>
        <scope>test</scope>
    </dependency>
</dependencies>
```

2. 配置项目

下面配置通用分页。这里我们使用的是Mybatis的Pagehelper插件。除了在Maven中引入pagehelper依赖外，还要引入Spring Boot自动配置的依赖：

```xml
<dependency>
    <groupId>com.github.pagehelper</groupId>
    <artifactId>pagehelper</artifactId>
    <version>5.1.2</version>
</dependency>
<dependency>
    <groupId>com.github.pagehelper</groupId>
    <artifactId>pagehelper-spring-boot-autoconfigure</artifactId>
    <version>1.2.3</version>
</dependency>
```

加入自动配置的依赖后，Spring Boot会自动配置拦截器等信息。

接下来我们修改application.properties文件的内容。配置系统访问的端口地址、数据库连接信息、MyBatis的mapper文件、分页拦截器以及POJO类扫描地址，另外配置日志打印信息。

```
server.port=8080

#数据库信息配置
spring.datasource.driver-class-name=com.mysql.cj.jdbc.Driver
spring.datasource.username=root
spring.datasource.password=123456
spring.datasource.url=jdbc:mysql://localhost:3306/hrms?useUnicode=true&characterEncoding=utf-8&useSSL=true&serverTimezone=UTC

#MyBatis相关配置
mybatis.mapper-locations=classpath:mapping/*Mapper.xml
mybatis.type-aliases-package=com.lhz.hrms.entity

#分页pageHelper
pagehelper.helper-dialect=mysql
pagehelper.reasonable=true
pagehelper.support-methods-arguments=true

#日志打印
logging.level.comlhz.mapper=debug
```

到此，Spring集成Spring MVC、MyBatis的环境已经基本搭建好了。

3. 引入前端框架LayUI

接下来我们引入前端框架LayUI。目前该框架的官网已停止维护，所有内容均已搬迁到Gitee上进行托管。LayUI的组件代码下载地址为https://gitee.com/sentsin/layui。Layui组件的详细使用教程可参考官方文档，地址是https://www.layui.com/doc/。下载完成后解压，将layui文件夹复制到项目的static路径下。完成后项目结构如图2.9所示。

然后在Spring Boot配置文件中配置静态文件访问方式。

```
# 配置静态文件访问方式
spring.mvc.static-path-pattern=/**
```

4. 配置JSP相关内容

本项目采用的是JSP动态网页技术，所以还要进行JSP相关的配置。首先引入JSP相关的依赖：

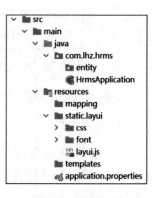

图2.9　项目结构

```xml
<!-- 添加servlet依赖模块 -->
<dependency>
    <groupId>javax.servlet</groupId>
    <artifactId>javax.servlet-api</artifactId>
</dependency>
<!-- 添加jstl标签库依赖模块 -->
<dependency>
    <groupId>javax.servlet</groupId>
    <artifactId>jstl</artifactId>
</dependency>
<!-- 使用jsp引擎，Spring Boot内置tomcat没有此依赖 -->
<dependency>
    <groupId>org.apache.tomcat.embed</groupId>
    <artifactId>tomcat-embed-jasper</artifactId>
</dependency>
```

然后在Spring Boot配置文件中配置JSP相关内容：

```
# web相关
spring.mvc.servlet.path=/
# 定位模板的目录
spring.mvc.view.prefix=/views
# 给返回的页面添加后缀名
spring.mvc.view.suffix=.jsp
```

至此，基本开发环境已搭建完成，接下来进入系统开发阶段。

2.2　项目前期准备

2.2.1　项目需求说明

本章涉及的员工管理系统是一个简单的、针对人事管理的系统，主要包含的功能有员工信息管理、部门管理、考勤管理和工资管理等。主要实现的功能有部门信息维护、员工信息维护、员工信息查询、员工考勤记录、员工工资记录等。

2.2.2 系统功能设计

根据员工管理系统要实现的主要功能，设计对应的功能列表。除了必要的登录注册功能外，主要包含四个模块，如图2.10所示。

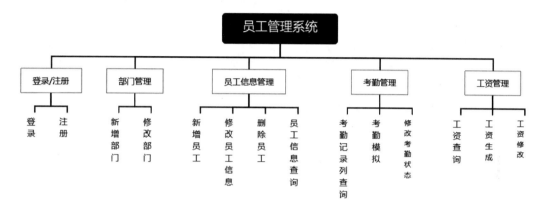

图2.10 员工管理系统结构图

2.2.3 系统数据库设计

（1）系统用户表，表名user，用于注册和登录。

```
CREATE TABLE 'user' (
  'id' int NOT NULL COMMENT '主键',
  'username' varchar(20) NOT NULL COMMENT '用户名',
  'realname' varchar(45) DEFAULT NULL COMMENT '真实姓名',
  'last_logine_time' datetime DEFAULT NULL COMMENT '上次登录时间',
  'createtime' datetime DEFAULT NULL COMMENT '注册/创建时间',
  'paword' varchar(32) NOT NULL COMMENT '密码',
  PRIMARY KEY ('id'),
  UNIQUE KEY 'username_UNIQUE' ('username'),
  UNIQUE KEY 'id_UNIQUE' ('id')
) ENGINE=InnoDB DEFAULT CHARSET=utf8 COMMENT='系统用户表'
```

（2）员工表，表名employee，是员工信息表。

```
CREATE TABLE 'employee' (
  'id' int NOT NULL AUTO_INCREMENT,
  'em_name' varchar(45) NOT NULL COMMENT '员工姓名',
  'birthday' varchar(45) NOT NULL COMMENT '出生日期',
  'sex' char(1) NOT NULL COMMENT '性别, F女, M男',
  'policital_status' int DEFAULT NULL COMMENT '政治面貌,0群众,1团员,2中共党员,3其他民主党派人士',
  'email' varchar(45) DEFAULT NULL COMMENT '邮箱',
  'education' int DEFAULT NULL COMMENT '学历,0无,1小学,2中学,3高中,4职高,5中专,6大专,7大专,8本科,9硕士,10博士',
  'university' varchar(45) DEFAULT NULL COMMENT '毕业院校',
  'station' varchar(45) DEFAULT NULL COMMENT '岗位',
```

```sql
  'salary_level' int DEFAULT NULL COMMENT '薪资级别',
  'emnum' varchar(45) NOT NULL COMMENT '工号',
  'dept' int NOT NULL COMMENT '所属部门',
  UNIQUE KEY 'id_UNIQUE' ('id'),
  UNIQUE KEY 'emnum_UNIQUE' ('emnum')
) ENGINE=InnoDB DEFAULT CHARSET=utf8 COMMENT='员工表'
```

（3）部门表，表名department，是部门信息表。

```sql
CREATE TABLE 'department' (
  'id' int NOT NULL AUTO_INCREMENT,
  'deptname' varchar(45) NOT NULL COMMENT '部门名称',
  PRIMARY KEY ('id'),
  UNIQUE KEY 'id_UNIQUE' ('id'),
  UNIQUE KEY 'deptname_UNIQUE' ('deptname')
) ENGINE=InnoDB DEFAULT CHARSET=utf8 COMMENT='部门表'
```

（4）考勤表，表名worktime，用来记录考勤信息。

```sql
CREATE TABLE 'worktime' (
  'id' int NOT NULL COMMENT '主键',
  'check_time' datetime NOT NULL COMMENT '考勤时间',
  'check_type' int NOT NULL COMMENT '0：上班考勤\n1：下班考勤\n',
  'status' int NOT NULL COMMENT '打卡状态，0：正常，1：迟到，2：早退，3：外勤',
  'employee_id' int NOT NULL COMMENT '对应员工id',
  PRIMARY KEY ('id'),
  UNIQUE KEY 'id_UNIQUE' ('id')
) ENGINE=InnoDB DEFAULT CHARSET=utf8 COMMENT='考勤表'
```

（5）工资表，表名salary，用于工资管理。

```sql
CREATE TABLE 'salary' (
  'id' int NOT NULL,
  'employee_id' int NOT NULL COMMENT '员工id',
  'base_salary' double NOT NULL COMMENT '基础工资',
  'merits_salary' double NOT NULL COMMENT '绩效',
  'award_salary' double NOT NULL COMMENT '奖金',
  'transport_subsidy' double NOT NULL COMMENT '交通补助',
  'meal_subsidy' double NOT NULL COMMENT '餐补',
  'phone_subsidy' double NOT NULL COMMENT '话费补助',
  'insurance' double NOT NULL COMMENT '保险费',
  'tax' double DEFAULT NULL COMMENT '个人所得税',
  'other_subsidy' double DEFAULT NULL COMMENT '其他加发',
  'other_cut' double DEFAULT NULL COMMENT '其他扣款',
  'belong_month' varchar(45) NOT NULL COMMENT '所属月份',
  PRIMARY KEY ('id'),
  UNIQUE KEY 'id_UNIQUE' ('id')
) ENGINE=InnoDB DEFAULT CHARSET=utf8 COMMENT='工资表'
```

2.2.4 系统文件说明

本项目系统文件的结构如图2.11所示。

下面对各个部分做简要说明：

（1）common：公用类，包含通用返回结果封装类、通用分页类和一些其他常用的公用类。

（2）config：代码配置类，Java配置类，配置一些不在配置文件中的信息，例如添加拦截器。

（3）controller：控制器类，所有controller层的类。

（4）实体类：entity是与数据库对应的实体类；qo是业务查询使用的类；vo是数据映射对象，用于页面展示。

（5）interceptor：拦截器类。

（6）mapper：使用MyBatis实现的数据访问层接口。

（7）service：业务逻辑的主要实现层，在该层可以实现事务。

（8）mapping：MyBatis映射文件，实现对数据访问层定义的接口。

（9）application.properties：Spring Boot的配置文件。

（10）static：静态资源，包含项目用到的JavaScript、CSS、图片、字体等文件，在本项目中引用的LayUI框架就放在这里。

（11）views：项目中所用到的所有的JSP页面。

（12）pom.xml：依赖JAR包，使用Maven进行项目构建时，依赖JAR包统一在pom文件中进行管理。

图2.11　项目文件组织结构

2.3　项目前端设计

2.3.1　登录注册

（1）登录

用户输入正确的用户名和密码，即可完成登录。登录页面如图2.12所示。

以下是登录页面的组件代码，基本都是使用LayUI提供的组件，详细的组件介绍可以参考LayUI的官方API文档。

图2.12　登录页面

```
<div class="login_main">
    <fieldset class="layui-elem-field layui-field-title hazel_mar_02">
        <legend>员工管理系统登录</legend>
    </fieldset>
    <div class="layui-row layui-col-space15">
        <form class="layui-form hazel_pad_01"
```

```html
                action="/login" method="post" enctype="application/json"
lay-filter="login">
            <div class="layui-col-sm12 layui-col-md12">
                <div class="layui-form-item">
                    <input type="text" name="username" lay-verify="required|
username" autocomplete="off" placeholder="账号" class="layui-input">
                    <i class="layui-icon layui-icon-username login_icon"></i>
                </div>
            </div>
            <div class="layui-col-sm12 layui-col-md12">
                <div class="layui-form-item">
                    <input type="password" name="paword" lay-verify="required|pass"
autocomplete="off" placeholder="密码" class="layui-input">
                    <i class="layui-icon layui-icon-password login_icon"></i>
                </div>
            </div>
            <div class="layui-col-sm12 layui-col-md12">
                <button class="layui-btn layui-btn-fluid" lay-submit
lay-filter="gofilter">登录</button>
            </div>
            <div class="layui-form-mid layui-word-aux"><a href="/registerPage">注
册</a></div>
            <div class="layui-form-mid layui-word-aux hint">
                <span id="loginmsg"></span>
            </div>
        </form>
    </div>
</div>
```

layui-form、layui-form-item等都是LayUI中常用的组件。layui-col-sm12、layui-col-md12等此类布局型的样式是LayUI自适应页面的实现。另外还有常用的图标layui-btn、按钮layui-btn等各种样式的组件。

以下是登录页面的部分JavaScript代码，用户输入账号和密码后调用按钮绑定的提交事件，向后台发送登录请求。

```javascript
layui.use('form', function () {//加载form模块
    var form = layui.form;
    var $ = layui.$;
    form.verify({
        username: function(value, item){ //value：表单的值；item：表单的DOM对象
            if(!new RegExp("^[a-zA-Z0-9_\u4e00-\u9fa5\\s·]+$").test(value)){
                return '用户名不能有特殊字符';
            }
            if(/(^\_)|(\__)|(\_+$)/.test(value)){
                return '用户名首尾不能出现下划线\'_\'';
            }
            if(/^\d+\d+\d$/.test(value)){
                return '用户名不能全为数字';
            }
        },
```

```
    //我们既支持上述函数式的方式，也支持下述数组的形式
    //数组的两个值代表[正则匹配，匹配不符时的提示文字]
    ,pass: [/^[\S]{6,12}$/,'密码必须6到12位，且不能出现空格']
});
//submit登录按钮
form.on('submit(gofilter)', function (data) {
    console.log(data);
    var user = {};//构造user对象，传递参数
    user.username = $("input[name=username]").val();
    user.paword = hex_md5($("input[name=paword]").val());//密码使用MD5加密
    $.ajax({
        url: "/login",
        type: "POST",
        data: JSON.stringify(user),//传递JSON类型的参数
        contentType: "application/json; charset=utf-8",
        dataType: "json",
        success: function (result) {
            if(result.code == COMMON_SUCCESS_CODE){
                location.href = "/index"// 登录成功跳转index页面
            }else{
                $("#loginmsg")[0].innerHTML = result.msg;//显示登录的错误信息
            }
        }
    })
    return false;//拦截LayUI自带的提交
})
});
```

提交表单前会对输入内容做验证，这里使用的是LayUI组件中的lay-verify属性，多种验证规则使用竖线（|）隔开。例如账号验证，对应的input输入框中lay-verify属性的值为"required|username"，表示该值为必填且需要通过username这个方法的验证，该方法定义在上面JavaScript代码中的form.verify方法中。密码的验证规则同理。

提交表单时使用了LayUI的过滤器form.on('submit(gofilter)'，对应的是HTML代码中登录按钮的属性lay-filter="gofilter"。发送请求使用的是JQuery中的ajax异步请求框架。请求发送成功后会调用success中定义的回调方法，解析返回结果，进行下一步处理，跳转至主页面或继续留在登录页面。

传递参数时，前端使用hex_md5方法对密码进行了加密传输。hex_md5是MD5加密算法中的常用算法。使用时将MD5.js文件引入statis文件夹即可。该文件是MD5算法JavaScript版本的实现。

（2）注册

在登录页面显示"注册"按钮，单击"注册"按钮跳转至注册页面，如图2.13所示。这里使用的机制比较简单，输入用户名、密码和确认密码即可完成注册。注册时校验两次输入的密码是否相同。

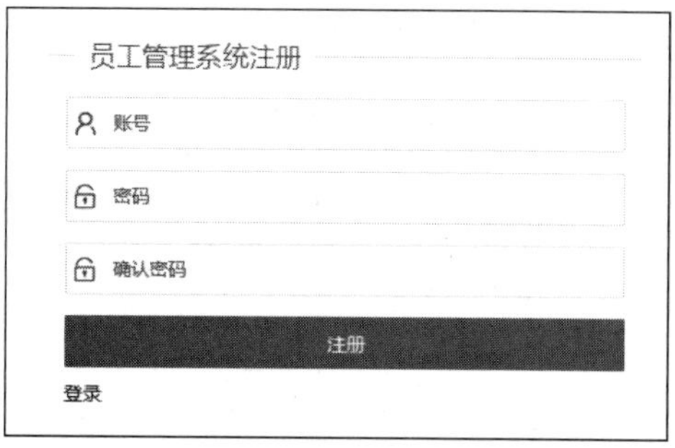

图2.13 注册页面

这部分的实现与登录基本相似，只是请求的后台接口路径以及请求返回后跳转的页面不同。注册时，单击"注册"按钮，绑定的是注册接口，接口成功返回后，根据返回结果跳转至登录页面或者继续留在注册页面。

2.3.2 部门管理

部门管理主要涉及部门的查询、增加和修改。部门信息是员工信息管理的基础信息。部门管理页面如图2.14所示。

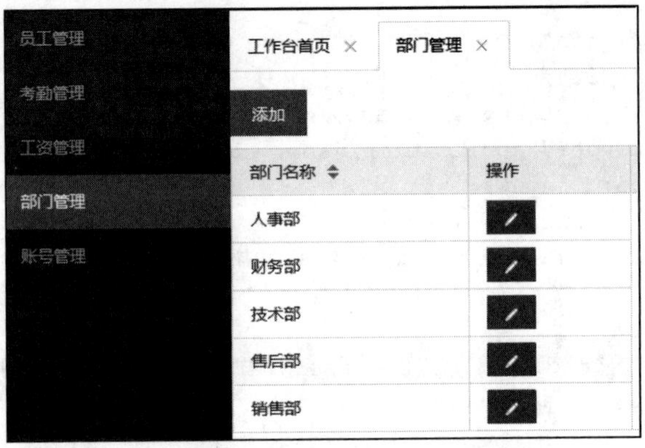

图2.14 部门管理页面

这里主要实现了部门的列表查询和添加功能，列表查询主要使用LayUI的table和button组件。HTML代码如下：

```html
<body>
    <div class="layui-btn-group layui-row" id="btn_group">
        <button type="button" class="layui-btn" data-method="addDept">添加</button>
    </div>
    <table class="layui-hide" id="deptTable"></table>
</body>
```

以上为部分HTML代码，定义table元素加载数据信息，定义button元素绑定添加操作。JavaScript代码如下：

```javascript
var deptTable = table.render({
    elem: '#deptTable'                    //table的id
    ,url:'/dept/list'                     //查询接口地址
    ,method:"post"
    ,data: JSON.stringify(dept)           //传递JSON类型的参数
    ,contentType: 'application/json'
    ,cellMinWidth: 80                     //全局定义常规单元格的最小宽度，LayUI 2.2.1新增
    ,cols: [[
        {field:'deptname', width:200, title: '部门名称', sort: true}
        ,{width:137, title: '操作',templet:function(d){
            return '<button type="button" class="layui-btn layui-btn-sm singleBtn" data-method="editDept" data-id="'+d.id+'"><i class="layui-icon">&#xe642;</i></button>'
```

```
            }}
        ]],
        parseData: function(res){          //res 即为原始返回的数据
            return res
        },
        done: function(res, curr, count){
            //如果是异步请求数据方式,res即为接口返回的信息
            //如果是直接赋值的方式,res即为:{data: [], count: 99}。data为当前页数据,count
为数据总长度
            $('.singleBtn').on('click', function(){      //绑定编辑按钮方法
                var othis = $(this), method = othis.data('method');
                active[method] ? active[method].call(this, othis) : '';
            });
        },
        page: false
    });
    //定义按钮触发事件
    var active = {
        addDept: function () {                    //添加部门页面弹窗
            var that = this;
            layerForm();                          //打开页面
        },
        editDept:function(){                      //编辑部门页面弹窗
            var othis = $(this), dataId = othis.data('id');
            $.ajax({
                url: "/dept/search/"+dataId,      //获取被编辑的部门信息
                type: "GET",
                contentType: "application/json; charset=utf-8",
                success: function (result) {
                    if(result.code == COMMON_SUCCESS_CODE){
                        var dept = result.data;
                        //给表单赋值
                        //addDeptForm即class="layui-form"所在元素属性lay-filter=""对应的值
                        form.val("addDeptForm", {
                            "deptname": dept.deptname  // "name": "value"
                                                ,"id": dept.id
                        });
                        layerForm();       //打开页面
                    }else{
                        layer.alert('数据获取失败', {icon: 2});
                    }
                }
            })
        }
    }
```

前端代码的按钮定义语句中,声明了属性data-method,值为"addDept",表示单击按钮时执行addDept方法。同理,部门编辑按钮中也绑定了编辑方法editDept。这些方法定义在JavaScript中。这两个方法打开的其实是同一个页面,不同点在于,打开编辑页面时,需要提前获取部门信息渲染到页面中,而添加页面是页面的初始状态。后面使用到的LayUI框架的按钮绑定方式基本和这里的处理方式类似。

JavaScript中定义的deptTable变量用来渲染表格数据，对应的是框架的render方法。该方法中elem属性为HTML代码中table元素的id值。cols属性用来对应结果集，在该属性中field的值与返回结果中字段名对应，即可自动获取该字段对应的值。如果想要对数据做些处理再显示，可以使用templet声明处理方法。例如部门管理表格中，操作列要显示编辑按钮，按钮中需要绑定这一行对应的部门id，那么就可以直接返回按钮对应的HTML代码。parseData是预处理数据，如果结果集中有些字段需要统一预处理，可以写在这里。

2.3.3 员工信息管理

员工管理主要涉及员工的添加、修改和删除。员工管理页面如图2.15所示。

图2.15 员工管理页面

以下为员工管理页面对应的部分代码。

```
var employeeTable = table.render({
    elem: '#employee'                          //页面元素id
    ,url:'/employee/list'                      //加载接口地址
    ,method:"post"                             //请求方式
    ,contentType: 'application/json'           //请求数据格式
    ,cellMinWidth: 80         //全局定义常规单元格的最小宽度，layui 2.2.1 新增
    ,cols: [[
        {field:'emnum', width:80, title: '工号', sort: true}
        ...
        templet:function(d){
            if(d.sex == "F"){
                return "女"
            }
            if(d.sex == "M"){
                return "男"
            }
        }]]
        ...
        ,{width:137, title: '操作',templet:function(d){
```

```
                return '<button type="button" class="layui-btn layui-btn-sm singleBtn"
data-method="editEmployee" data-id="'+d.id+'"><i class="layui-icon">&#xe642;
                </i></button>' +' <button type="button" class="layui-btn
layui-btn-sm singleBtn" data-method="deleteEmployee" data-id="'+d.id+'"><i class=
"layui-icon">&#xe640; </i></button>'
            }}
        ]],
        ...
        page: true
    });
    //删除按钮绑定事件
    deleteEmployee:function(){
        var othis = $(this), dataId = othis.data('id');              //得到员工id
        layer.confirm('确定删除？', {
            btn: ['确定', '取消']                                      //可以有无限个按钮
            ,yes: function(index, layero){
                var layDelete = layer;
                $.ajax({
                    url: "/employee/delete/"+dataId,                 //删除接口地址
                    type: "DELETE",                                  //请求方式
                    contentType: "application/json; charset=utf-8",  //请求内容格式
                    success: function (result) {
                        if(result.code == COMMON_SUCCESS_CODE){
                            employeeTable.reload({page:{curr:1}})    //重载表格数据
                        }else{
                            layer.alert('删除失败', {icon: 2});
                        }
                        layDelete.closeAll();
                    }
                })
            }, btn2: function(index, layero){
                layer.closeAll();
            }
        });
    }
```

性别这一列的表头使用了数据转换，在templet变量中定义转换函数。这里将返回的性别字符转换成了对应的中文，男或女。

操作列新增了一个删除按钮，与编辑按钮同理，不同点在于，删除按钮不需要打开页面，只需要弹出一个提示框，根据用户的选择决定是否删除。提示框是LayUI内置的组件。LayUI有很多种类的提示框，使用时可以根据具体场景选择合适的提示框。

删除成功后，调用回调方法。如果成功，则使用table的reload方法重新刷新表格数据；如果失败，则提示失败。

添加员工时除填写基本信息外，还要选择所在部门，获取对应的部门列表，如图2.16所示。

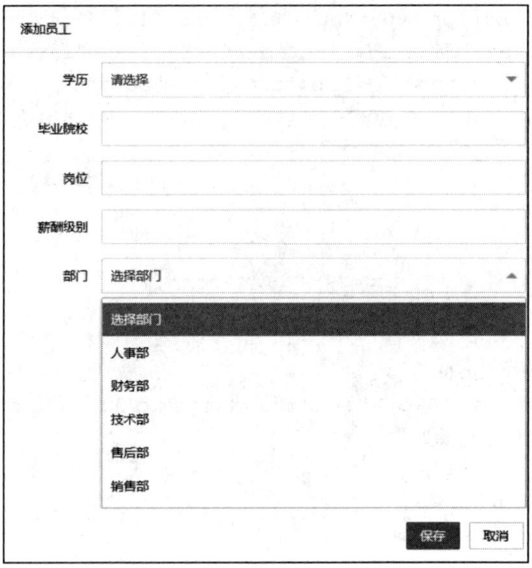

图2.16　添加员工

该页面对应的部分HTML代码如下：

```
<div hidden id="addEmployee">
    <form class="layui-form" action="" lay-filter="addEmployeeForm" id="addEmployeeForm">
        ...
        <div class="layui-form-item">
            <label class="layui-form-label">部门</label>
            <div class="layui-input-block">
                <select name="dept" lay-filter="dept" class="dept">
                    <option value="">选择部门</option>
                </select>
            </div>
        </div>
    </form>
</div>
```

代码中使用了LayUI的select组件，这是常用的下拉框组件。其中部门列表的下拉框属于异步加载，异步加载的JavaScript代码如下：

```
//获取部门列表
$.ajax({
    url: "/dept/list",
    type: "POST",
    data: JSON.stringify(dept),//传递JSON类型的参数
    contentType: "application/json; charset=utf-8",
    dataType: "json",
    success: function (result) {
        if(result.code == COMMON_SUCCESS_CODE){
            $.each(result.data,function(index,item){
                $('.dept').append(new Option(item.deptname,item.id));//往下拉菜单里添加元素
            })
```

```
            form.render();//菜单渲染，把内容加载进去
        }else{
            layer.alert('获取部门信息失败', {icon: 2});
        }
    }
})
```

这里在请求到部门数据后，将部门数据先添加到form元素中，然后调用form的render方法实现最终数据渲染。

对于页面上需要请求后台接口的数据，尽量使用异步加载。异步加载可使用在有多请求的页面，页面渲染和数据请求同时进行，这样可以缩短页面打开的时间，提高页面渲染效率，提升用户使用体验。默认正常模式下，浏览器使用同步加载，也叫阻塞加载模式，这种模式类似瀑布模式，按顺序执行操作，因此停止了后续的文件加载（如图像）、渲染、代码执行等。异步加载又叫非阻塞加载模式，是一种同步执行的模式。同步加载和异步加载不只是在初始加载页面时有区别，在刷新数据时，同步加载会将整个页面数据都重新加载，而异步加载只需要加载被刷新的部分。

异步加载的优点在于：

（1）可以同时从服务器请求多项内容

如果页面中存在多个图片或其他耗时文件，或者需要加载多项数据，使用异步模式会大大提高效率。

（2）提高请求返回速度

由于每个请求都只加载特定的内容，因此数据量小时，请求速度会相应提高。由于多项请求同时进行，时间并非累加，而是并行，所以请求时间会缩短。

（3）只改变需要改变的数据，能够减少服务器数据流量

对于一些需要刷新的页面，使用异步模式可以只刷新特定的部分，而不需要重新加载整个页面，这样能够减少流量交互，提高响应速度。

（4）可以进行其他操作

发送请求后，在等待请求响应的过程中，用户可以进行其他操作。

常见的异步加载形式：

（1）dom加载JavaScript文件

默认情况下，使用如下方式加载JavaScript文件：

```
<script src="http:/XXX/script.js"></script>
```

改为使用dom异步加载JavaScript文件：

```
(function() {
    var s = document.createElement('script');              //获取dom元素
    s.type = 'text/javascript';                            //设置文件类型
    s.async = true;                                        //设置异步
    s.src = 'http://XXX/script.js';                        //文件地址
    var x = document.getElementsByTagName('script')[0];    //定义插入位置
```

```
        x.parentNode.insertBefore(s, x);
})();
```

这样在加载JavaScript文件的时候不影响页面的其他操作。

(2) Ajax异步请求

例如前面的添加员工页面,在打开页面后,使用Ajax发送异步请求返回部门列表,渲染到页面中。

(3) Onload异步加载

例如前面的JavaScript文件加载,也可以改为onload形式来加载,代码如下:

```
(function() {
    function load(){
        var s = document.createElement('script');            //获取dom元素
        s.type = 'text/javascript';                          //设置文件类型
        s.async = true;                                      //设置异步
        s.src = 'http://XXX/script.js';                      //文件地址
        var x = document.getElementsByTagName('script')[0];  //定义插入位置
        x.parentNode.insertBefore(s, x);
    }
    window.attachEvent('onload', async_load);                //绑定事件
})();
```

添加员工时弹出添加页面,弹窗设置如下:

```
function layerForm(){
    //多窗口模式,层叠置顶
    layer.open({
        type: 1
        , title: '添加员工'
        , content: $('#addEmployee')                         //页面内容div的id
        , btn: ['保存', '取消']                               //只是为了演示
        , area: ['600px', '600px']                           //宽和高
        ,closeBtn: 0
        , yes: function () {
            var formData = form.val('addEmployeeForm');      //对应form的id
            var layerui = layer;                             //使用layer组件
            $.ajax({
                url: "/employee/add",                        //添加接口地址
                type: "POST",
                data: JSON.stringify(formData),              //传递JSON类型的参数
                contentType: "application/json; charset=utf-8",
                dataType: "json",
                success: function (result) {
                    if(result.code == COMMON_SUCCESS_CODE){
                        layerui.msg('操作成功');
                        employeeTable.reload({page:{curr:1}}) //刷新数据
                    }else{
                        layerui.alert('操作失败', {icon: 2}); //icon的数字定义可参考
                                                              LayUI官方文档
                    }
```

```
                    layerui.closeAll();                    //关闭弹窗
                    $("#addEmployeeForm")[0].reset();//页面数据重置,防止下次打开时遗留数据
                    form.render();                         //重置数据渲染
                }
            })
        }
        , btn2: function () {                              //取消按钮回调
            layer.closeAll();                              //关闭弹窗
            $("#addEmployeeForm")[0].reset();
            form.render();
        }
    });
}
```

弹窗使用的是LayUI的layer组件。layer组件可以实现弹窗的各项设置,包括页面的布局、标题、按钮以及各按钮的回调方法。content对应的是在html中定义的容器id,该容器存放的是整个弹窗要显示的内容。btn是显示的按钮文本,按顺序展示。area定义弹窗的宽高。yes是保存按钮的回调方法,在该方法中使用Ajax异步发送添加员工的请求,并在请求成功后执行回调方法。btn2定义了取消按钮的回调方法。

2.3.4 工资管理

工资管理包括工资的生成和修改。生成时输入每一项对应的数额,然后合计出总数。计算公式可以放在前端页面,也可以在后台生成后返回前端。这里采用简单的方式,统一由前端计算。工资页面如图2.17所示。

图2.17 工资管理页面

在工资管理页面上方的搜索栏中有选择工资月份选项,这里用到了LayUI中的Laydate组件,代码如下:

```
var laydate = layui.laydate;
var dept = {};
//日期
laydate.render({
    elem: '#belongMonth'            //日期元素id
    ,format: 'yyyy-MM'              //日期格式
    ,type: 'month'                  //类型
});
```

laydate组件包含年选择器、年月选择器、日期选择器、时间选择器、日期时间选择器,这5种类型的选择器均支持范围选择(即双控件),内置强劲的自定义日期格式解析和合法校正机制。由于其内部采用的是零依赖的原生JavaScript编写,因此又可作为独立组件使用。laydate的使用方式如表2.1所示。

表2.1 laydate的使用方式

场 景	用前准备	调用方式
在layui模块中使用	下载LayUI后,引入layui.css和layui.js即可	通过layui.use('laydate', callback)加载模块后,再调用方法
作为独立组件使用	下载独立组件包,引入laydate.js即可	直接调用方法使用

计算公式可写在templet属性中,定义好对应的function函数即可。这里只进行了简单的加减运算,读者可以加大难度,练习复杂数学函数计算。代码如下:

```
var salaryTable = table.render({
   ...
   ,{field:'count', width:100, title: '合计',templet:function(d){
      return d.baseSalary+d.meritsSalary+d.awardSalary+
         d.transportSubsidy+d.mealSubsidy+d.phoneSubsidy+d.otherSubsidy
         -d.insurance-d.tax-d.otherCut;
   }}
   ...
});
```

生成工资页面如图2.18所示。生成工资时,填入基本参数,选择部门和员工即可。

图2.18 生成工资页面

生成工资页面部分代码如下:

```
<div hidden id="generateSalary">
   <form class="layui-form" action="" lay-filter="generateSalaryForm" id="generateSalaryForm">
      <div class="layui-form-item" id="imitateDeptDiv" >
         <label class="layui-form-label">部门</label>
         <div class="layui-input-block">
```

```html
            <select name="dept" lay-filter="dept" class="dept" id="imitateDept" lay-verify="required" >
                <option value="">选择部门</option>
            </select>
        </div>
    </div>
    <div class="layui-form-item"  id="employeeGenerateDiv">
        <label class="layui-form-label">员工</label>
        <div class="layui-input-block">
            <select name="employeeId" lay-filter="employeeId" class="employeeImitate" lay-verify="required">
                <option value="">选择员工</option>
            </select>
        </div>
    </div>
    ...
</div>
```

与前面异步获取部门不同的是，这里的部门和员工下拉框需要联动，先使用异步加载功能加载部门列表，选择了部门之后，动态加载部门下的员工列表。

```javascript
form.on('select(dept)', function(data){
    var selectdept = data.elem;
    if(data.value != null){
        var employee={};
        employee.dept = data.value;
        $.ajax({
            url: "/employee/searchAll",
            type: "POST",
            data: JSON.stringify(employee),              //传递JSON类型的参数
            contentType: "application/json; charset=utf-8",
            dataType: "json",
            success: function (result) {
                if (result.code == COMMON_SUCCESS_CODE) {
                    if(selectdept.id == 'searchDept'){
                        $(".employee option:gt(0)").remove();   //移除之前选择的员工
                        $.each(result.data, function (index, item) {
                            $('.employee').append(new Option(item.emName, item.id));
                                                //往下拉列表里添加员工
                        })
                    }else{
                        $(".employeeImitate option:gt(0)").remove();//移除之前选择的员工
                        $.each(result.data, function (index, item) {
                            $('.employeeImitate').append(new Option(item.emName, item.id));//往下拉列表里添加员工
                        })
                    }
                    form.render();              //渲染，把内容加载进去
                } else {
```

```
                layer.alert('获取信息失败', {icon: 2});
            }
        }
    });
}
});
```

使用select元素的lay-filter属性进行过滤，部门的filter值设置为dept，然后在JavaScript中使用from元素绑定下拉框选择事件select(dept)。这样当选择部门后或者部门发生改变后，会动态发送异步请求到后台获取该部门下的员工列表，并将数据渲染到员工select元素中。

注意：当部门发生改变时，要及时清除之前选择的员工，避免数据展示错误。退出当前页面时，也要记得重置页面数据。

2.3.5 考勤记录管理

考勤记录管理包括员工的打卡记录。这里需要对接移动端，可以使用小程序或者手机App。由于篇幅有限，这里不再详述。本项目在后台模拟了打卡操作，实际应用中可以使用移动端调用该接口，原理相同。考勤记录管理页面如图2.19所示。

图2.19 考勤记录管理页面

使用页面模拟接口调用进行打卡的页面如图2.20所示。

模拟打卡页面代码如下：

```
layer.open({
    type: 1
    , title: '模拟打卡--'
    , content: $('#imitateCheck')
    , btn: ['打卡', '取消']  //只是为了演示
    , area: ['600px', '600px']  //宽和高
    ,closeBtn: 0
    , yes: function () {
```

```
            var formData = form.val('imitateCheckForm');
            console.log(formData);
            var layerui = layer;
            $.ajax({
                url: "/worktime/imitateCheck",//模拟打卡接口地址
                type: "POST",
                data: JSON.stringify(formData),//传递JSON类型的参数
                contentType: "application/json; charset=utf-8",
                dataType: "json",
                success: function (result) {
                    if(result.code == COMMON_SUCCESS_CODE){//模拟成功
                        layerui.msg('操作成功');
                        workTable.reload({page:{curr:1}})
                    }
...
```

实际应用中，此功能应该只提供接口，供打卡客户端调用即可。

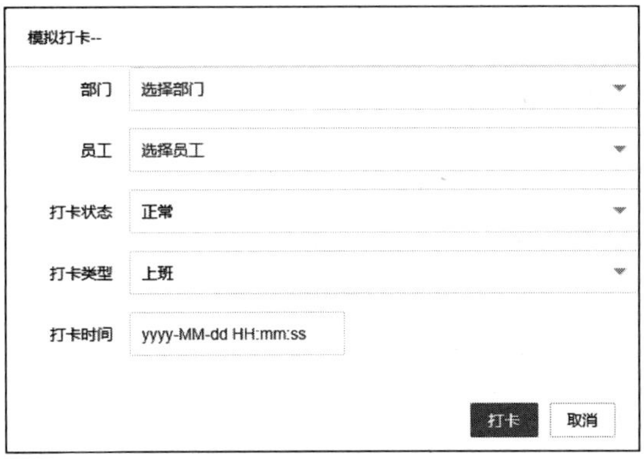

图2.20　模拟打卡

2.4　项目后端实现

2.4.1　通用分页类

1. 分页参数传递与结果封装

本项目中所有使用分页的需求,查询时的参数统一继承自PageBean分页类。设置通用分页,可以避免代码冗余,也便于和前端接口进行对接,防止出现不同接口分页参数命名不同的情况。

```
//分页类
public class PageBean {
    private int limit=10;     //每页数量
    private int page=1;       //当前页面
    public int getLimit() {
        return limit;
    }
```

```java
    public void setLimit(int limit) {
        this.limit = limit;
    }
    public int getPage() {
        return page;
    }
    public void setPage(int page) {
        this.page = page;
    }
}
//分页返回结果封装类
public class PageCommonResult extends CommonResult{
    private long count;//总条数
    public long getCount() {
        return count;
    }
    public void setCount(long count) {
        this.count = count;
    }
    public void setCount(int count) {
        this.count = count;
    }
    public static  <T extends Object> PageCommonResult  success(List<T> data, long count){
        PageCommonResult pageCommonResult = new PageCommonResult();
        pageCommonResult.setCount(count);                            //设置总条数
        pageCommonResult.setCode(CommonCode.SUCCESS);                //设置状态码
        pageCommonResult.setData(data);                              //设置返回数据列表
        return pageCommonResult;
    }
    public PageCommonResult(){};
}
```

2. PageHelper分页实现原理

首先我们看一下如何使用**PageHelper**实现分页查询，以下是一个简单的示例。

```java
public PageCommonResult employeeList(@RequestBody EmployeeQO employee){
    PageHelper.startPage(employee.getPage(),employee.getLimit());  //设置分页参数
    List<EmployeeVO> employeeList = employeeService.list(employee);//执行查询
    PageInfo<EmployeeVO> pageInfo = new PageInfo(employeeList);    //封装分页结果
    //返回结果
    PageCommonResult commonResult = PageCommonResult.success(pageInfo.getList(),pageInfo.getTotal());//设置通用返回结果
    return commonResult;
}
```

查询参数传递过来的对象EmployeeQO继承了前面讲述的PageBean，所以里面包含了分页查询相关的参数。首先使用PageHelper对象调用startPage方法设置分页参数，设置完之后执行查询语句。由于之前在Spring Boot配置文件中配置了PageHelper，所以这里会自动封装分页SQL语句，而不需要开发者手动设置。执行完毕，将数据封装到通用返回结果中即可。

接下来我们看一下PageHelper是如何通过这么简单的设置来实现分页的。

首先我们要知道MyBatis的插件是如何正常引入并工作的。在MyBatis中插件是通过拦截器来实现的，所拦截的对象有Executor、StatementHandler、ParameterHandler、ResultSetHandler。这四种对象也是MyBatis实现SQL的重要插件。默认情况下，MyBatis允许使用插件来拦截的方法调用如表2.2所示。

表2.2　MyBatis允许使用插件来拦截的方法调用

拦截对象	方　　法	说　　明
Executor	update	更新
	query	查询
	flushStatements	刷新语句
	commit	提交事务
	rollback	回滚事务
	getTransaction	获取事务对象
	close	关闭执行器
	isClosed	获取执行器的关闭状态
ParameterHandler	getParameterObject	获取参数对象
	setParameters	设置参数对象
ResultSetHandler	handleResultSets	处理结果集
	handleOutputParameters	处理存储过程的输出参数
StatementHandler	prepare	预编译SQL
	parameterize	设置参数
	batch	批量操作
	update	执行更新操作的SQL语句
	query	执行查询操作的SQL语句

MyBatis使用插件拦截方法的流程如图2.21所示。

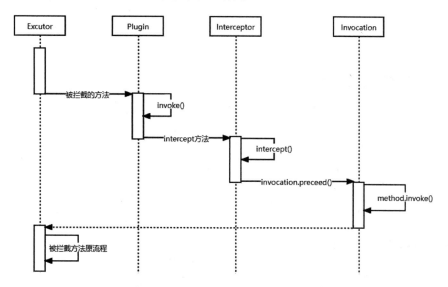

图2.21　MyBatis使用插件拦截方法的流程

这些类中方法的细节可以通过查看每个方法的签名来发现，或者直接查看MyBatis发行包中的源代码。这些都是更底层的类和方法，所以使用插件的时候要特别小心。通过MyBatis提供的强大机制，使用插件是非常简单的，只需实现Interceptor接口并指定想要拦截的方法签名即可。

PageHelper插件也是使用这个流程实现拦截的。PageHelper插件拦截的方法是Executor中的query方法。我们可以在它的拦截器中查看。从下面的代码中可以看到PageInterceptor实现了Interceptor接口，并且在class上添加了@Signature注解标明要拦截的方法签名。

```java
@Intercepts(
    {
        @Signature(type = Executor.class, method = "query", args =
{MappedStatement.class, Object.class, RowBounds.class, ResultHandler.class}),
        @Signature(type = Executor.class, method = "query", args =
{MappedStatement.class, Object.class, RowBounds.class, ResultHandler.class,
CacheKey.class, BoundSql.class}),
    }
)
public class PageInterceptor implements Interceptor {
    @Override
    public Object intercept(Invocation invocation) throws Throwable {
    }
    ...
}
```

然后Spring Boot会自动将这个拦截器配置到MyBatis的SessionFactory中，这样就可以实现query方法的拦截。

```java
@Configuration
@ConditionalOnBean(SqlSessionFactory.class)
@EnableConfigurationProperties(PageHelperProperties.class)
@AutoConfigureAfter(MybatisAutoConfiguration.class)
public class PageHelperAutoConfiguration {
    @Autowired
    private List<SqlSessionFactory> sqlSessionFactoryList;
    @Autowired
    private PageHelperProperties properties;
    /**
     * 接收分页插件额外的属性         *
     * @return
     */
    @Bean
    @ConfigurationProperties(prefix = PageHelperProperties.PAGEHELPER_PREFIX)
    public Properties pageHelperProperties() {
        return new Properties();
    }
    @PostConstruct
    public void addPageInterceptor() {
        PageInterceptor interceptor = new PageInterceptor();
        Properties properties = new Properties();
        //先把一般方式配置的属性放进去
```

```
            properties.putAll(pageHelperProperties());
            //再把特殊配置放进去，由于close-conn在利用上面方式时属性名就是 close-conn 而不是
closeConn, 所以需要额外的一步
            properties.putAll(this.properties.getProperties());
            interceptor.setProperties(properties);
            for (SqlSessionFactory sqlSessionFactory : sqlSessionFactoryList) {
                //加入拦截器
                sqlSessionFactory.getConfiguration().addInterceptor(interceptor);
            }
        }
    }
```

被拦截的方法会先执行上述PageInterceptor拦截器中的intercept方法，为SQL语句设置分页参数，然后执行Invocation中的proceed()方法，该方法内会执行method.invoke()方法，这样就可以继续执行被拦截方法原来的逻辑。这就是图2.21中所展示的整个流程。

2.4.2 通用返回结果

返回结果使用CommonResult统一封装，便于前端解析，也便于后端开发，提高效率，返回结果包括常用的返回结果封装形式，例如成功结果封装、失败结果封装、分页结果封装、数据结果封装等。使用时直接在相关位置调用即可。

```
//方法执行成功后，返回数据对象时使用
public static <T extends Object> CommonResult success(Object data) {
    CommonResult commonResult = new CommonResult();
    commonResult.setCode(CommonCode.SUCCESS);            //设置成功码
    commonResult.setData(data);                          //设置返回数据
    return commonResult;
}
//方法执行失败后返回提示信息
public static CommonResult fail(String msg) {
    CommonResult commonResult = new CommonResult();
    commonResult.setCode(CommonCode.FAIL);               //设置失败码
    commonResult.setMsg(msg);                            //设置提示消息
    return commonResult;
}
```

2.4.3 登录/注册

登录/注册的流程如图2.22所示。

1. 登录

接收到前端请求的用户名和密码后，与数据库中的用户作校验，返回登录结果。登录完成后将用户放入session，用于保持登录和拦截器校验。

以下是登录功能的部分代码。

```
CommonResult commonResult = new CommonResult();
//1. 校验参数
    if(user == null || StringUtils.isEmpty(user.getUsername())||StringUtils.isEmpty(user.getPaword())){
```

```java
        commonResult.setCode(CommonCode.FAIL);
        commonResult.setMsg("用户名和密码不能为空");
        return commonResult;
    }
    //2. 根据用户名查询用户
    User userByUsername = userService.getUserByUsername(user.getUsername());
    if(userByUsername!=null){
        //2-1. 成功,将用户信息放入session,更新最新登录时间
        if(userByUsername.getPaword().equals(user.getPaword())){
            //更新最新登录时间
            userByUsername.setLastLogineTime(new Date());
            userService.edit(userByUsername);
            commonResult.setCode(CommonCode.SUCCESS);
            request.getSession().setAttribute(CommonCode.SESSION_USER,user);
        }else{//2-2. 失败,提示失败信息
            commonResult.setCode(CommonCode.FAIL);
            commonResult.setMsg("密码不正确");
        }
    }else{
        commonResult.setCode(CommonCode.FAIL);
        commonResult.setMsg("用户名不正确");
    }
    return commonResult;
```

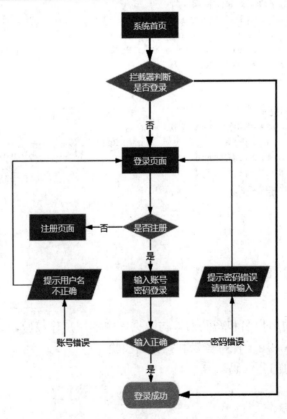

图2.22 登录/注册流程图

2. 注册

以下是注册相关的部分代码。

```
CommonResult commonResult = new CommonResult();
//1. 校验参数
if(user == null || 
StringUtils.isEmpty(user.getUsername())||StringUtils.isEmpty(user.getPaword())){
    commonResult.setCode(CommonCode.FAIL);
    commonResult.setMsg("用户名和密码不能为空");
    return commonResult;
}
//2. 根据用户名查找，如果存在，提示已存在
User userByUsername = userService.getUserByUsername(user.getUsername());
if(userByUsername!=null){
    commonResult.setCode(CommonCode.FAIL);
    commonResult.setMsg("用户名已存在");
}else{//3. 如果不存在，存储加密后的用户信息
    //设置创建时间，新增用户
    user.setCreatetime(new Date());
    int result = userService.addUser(user);
    if(result == 0){
        commonResult.setCode(CommonCode.SUCCESS);
    }else{
        commonResult.setCode(CommonCode.FAIL);
        commonResult.setMsg("系统原因，请重试或联系管理员");
    }
}
return commonResult;
```

注册时校验用户名是否合法、用户名是否存在、密码是否合法。注册后将使用MD5算法把加密后的密码存入数据库中，不保存原始密码。这是为了防止数据库信息泄露后造成用户账号信息泄露。

MD5全称是Message-Digest Algorithm，是一种常用的信息摘要算法，使用哈希算法函数进行加密，用于确保信息传输完整一致。它是一个安全的散列算法，输入两个不同的明文不会得到相同的输出值，其过程不可逆，所以无法解密，只能用穷举法把可能出现的明文用MD5算法散列后，把得到的散列值和原始的数据形成一个一对一的映射表，对比散列值从映射表中找出破解密码所对应的原始明文。MD5算法有以下几个特点：

（1）一致性：任意长度的数据得到的MD5值长度都是固定的。

（2）简便性：很容易从原数据计算出MD5值。

（3）抗修改性：对原数据进行任何改动，就算只修改1个字节，所得到的MD5值也不相同。

（4）弱抗碰撞：伪造数据非常困难。

（5）强抗碰撞：几乎无法实现两个不同的数据得到相同的MD5值。

目前它的主要应用途径有以下几种：

（1）密码加密

对用户账号的密码进行MD5加密后再存储到数据库中，可以避免用户账号信息泄露，提高了系统安全性。

（2）文件校验

文件存储时，可以使用MD5进行校验，这样可以判断文件是否完整、来源是否可靠等。

（3）请求参数校验

对请求进行参数校验可以避免接口恶意调用。常见的是使用用户账号相关信息与时间戳，通过MD5算法生成签名，与服务端进行校验。只要签名不同，不管请求参数如何正确，都将视为无效请求而被拦截器拦截。

3. 登录拦截器

编写对应的拦截器，用户未登录时无法访问系统，将跳转到登录页面。用户已登录的，继续执行用户操作。代码如下：

```java
/**
 * 在请求处理之前进行调用（Controller方法调用之前）
 */
@Override
public boolean preHandle(HttpServletRequest request, HttpServletResponse response, Object handler) {
    try {
        //统一拦截（查询当前session是否存在user, user会在每次登陆成功后写入session）
        User user=(User)request.getSession().getAttribute(CommonCode.SESSION_USER);
        if(user!=null){
            return true;
        }
        response.sendRedirect(request.getContextPath()+"/loginPage");
    } catch (IOException e) {
        e.printStackTrace();
    }
    return false;
}
```

拦截器编写完成后要在Spring中注册。可以另行编写配置文件，添加@Configuration注解。另外要排除不需要拦截的路径。

```java
@Configuration
public class HrmsLoginConfig implements WebMvcConfigurer {
    @Override
    public void addInterceptors(InterceptorRegistry registry) {
        //注册TestInterceptor拦截器
        InterceptorRegistration registration = registry.addInterceptor(new UserIntercepter());
        registration.addPathPatterns("/**");              //所有路径都被拦截
        registration.excludePathPatterns(                 //添加不拦截路径
```

```
            "/loginPage",                    //登录
            "/login",                        //登录
            "/registerPage",                 //注册
            "/register",                     //注册
            "/**/*.html",                    //HTML静态资源
            "/**/*.js",                      //JS静态资源
            "/**/*.css",                     //CSS静态资源
            "/**/*.woff",
            "/**/*.ttf"
        );
    }
}
```

2.4.4 部门管理

1. 部门查询

部门查询功能的代码如下：

```
/**
 * 部门查询不分页
 * @return
 */
@ResponseBody
@ApiOperation("部门查询不分页")
@RequestMapping(value = "/dept/list",method = RequestMethod.POST)
public CommonResult deptList(@RequestBody Department dept){
    List<DepartmentVO> depts = departmentService.list(dept);
    //返回结果
    CommonResult commonResult = CommonResult.success(depts);
    return commonResult;
}
```

因为部门信息相对较少，所以未使用分页查询，将查询结果使用通用返回结果进行封装。这里使用了@RequestMapping、@ResponseBody、@RequestBody注解。@RequestMapping注解用于实现地址映射。

Spring MVC提供多个注解实现接口地址映射，常用的有@RequestMapping、@PostMapping、@GetMapping、@PutMapping、@DeleteMapping。它们之间的关系和区别如表2.3所示。

表2.3 Spring MVC提供的实现接口地址映射的注解

注 解 名	说　　明	说　　明
@RequestMapping	需设置路径和请求方法，可用其他注解进行替代	例如@RequestMapping(value = "XXX",method = RequestMethod.POST)等同于@PostMapping(value="XXX")，另外Put、Delete、Get方法也同样能转化
@PostMapping	指定Post请求	@PostMapping(value="XXX")
@GetMapping	指定Get请求	@GetMapping(value="XXX")

（续表）

注 解 名	说　　明	说　　明
@PutMapping	指定Put请求	@PutMapping(value="XXX")
@DeleteMapping	指定Delete请求	@DeleteMapping(value="XXX")

@ResponseBody注解将方法的返回值以特定的格式写入response的body区域，进而将数据返回给客户端。若方法上面没有写ResponseBody,底层会将方法的返回值封装为ModelAndView对象。

@RequestBody注解主要用来接收前端传递给后端的请求体中的数据，常用于Post和Put请求。Get请求中常用的接收参数的注解是@RequestParam。

2. 部门添加与修改

部门添加与修改功能的代码如下：

```java
/**
 * 部门添加
 * @return
 */
@ResponseBody
@ApiOperation("部门添加")
@RequestMapping(value = "/dept/add",method = RequestMethod.POST)
public CommonResult deptAdd(@RequestBody Department dept){
    departmentService.add(dept);            //执行添加
    //返回结果
    CommonResult commonResult = CommonResult.success();
    return commonResult;
}
// service方法
@Override
public int add(Department dept) {
    if(dept.getId() !=null){
        return departmentMapper.updateByPrimaryKey(dept);
    }else{
        return departmentMapper.insert(dept);
    }
}
```

在Controller中部门添加与修改为同一方法。在service方法中通过校验有无id来判断是添加还是修改操作。如果严格按照接口规则来定义的话，可以将添加方法和修改方法拆开，添加接口的请求方法为Post，修改接口的请求方法为Put。本项目因为业务简单，因此将添加与修改都写在同一个接口，读者可以自行练习使用Put请求方法进行修改操作。

service方法直接调用Mapper提供的接口就可以实现数据库操作。Mapper提供的接口在mapper.xml文件中有对应的SQL实现语句。

```xml
<insert id="insert" parameterType="com.lhz.hrms.entity.Department">
    insert into department (
        id,
        deptname,
```

```
            chargeman
        )
    values (
            #{id,jdbcType=INTEGER},
            #{deptname,jdbcType=VARCHAR},
            #{chargeman,jdbcType=INTEGER}
        )
</insert>
```

其他业务实现逻辑与这个基本相同。

2.4.5 员工信息管理

（1）员工查询

员工查询功能代码如下：

```
/**
 * 员工查询分页
 * @return
 */
@ResponseBody
@ApiOperation("员工查询")
@RequestMapping(value = "/employee/list",method = RequestMethod.POST)
public PageCommonResult employeeList(@RequestBody EmployeeQO employee){
    PageHelper.startPage(employee.getPage(),employee.getLimit());  //设置分页参数
    List<EmployeeVO> employeeList = employeeService.list(employee); //数据列表
    PageInfo<EmployeeVO> pageInfo = new PageInfo(employeeList);     //分页信息
    //返回结果
    PageCommonResult commonResult = PageCommonResult.success(pageInfo.getList(),
pageInfo.getTotal());
    return commonResult;
}
/**
 * 单个员工详细信息查询
 * @return
 */
@ResponseBody
@ApiOperation("员工查询")
@RequestMapping(value = "/employee/search/{id}",method = RequestMethod.GET)
public CommonResult employeeSearch(@PathVariable Integer id){
    Employee employee = employeeService.findById(id);
    CommonResult commonResult = CommonResult.success(employee);
    return commonResult;
}
```

除了员工查询外，还有单个员工信息的展示。这里的分页查询使用的就是前面配置的 Pagehelper 插件。首先调用 startPage 方法设置分页参数，然后执行查询即可。得到对应的 list 结果集后，使用 PageInfo 类对结果进行分页转换。

单个查询使用的请求方法是 Get，但是这里没有使用 @RequestParam 注解，而是使用了路径通配符，将参数直接在路径中进行传递。这里用到了 @PathVariable 注解，默认情况下该注

解对应的参数名与路径中声明的通配符相同,如果不相同,可以在注解中进行指定,指定时需要保证注解内声明的参数名和通配符一致。例如@PathVariable("emId"),其路径应为"/employee/search/{emId}"。

(2)员工添加或修改

员工添加或修改功能的代码如下:

```
/**
 * 员工添加或修改
 * @return
 */
@ResponseBody
@ApiOperation("员工添加或修改")
@RequestMapping(value = "/employee/add",method = RequestMethod.POST)
public CommonResult employeeAdd(@RequestBody Employee employee){
    if(employee!=null && employee.getId()!=null){
        employeeService.edit(employee);
    }else{
        employeeService.add(employee);
    }
    //返回结果
    CommonResult commonResult = CommonResult.success();
    return commonResult;
}
```

校验参数id确定是添加还是修改员工。添加员工调用add()方法,修改员工调用edit方法。

(3)员工删除

员工删除功能的代码如下:

```
/**
 * 员工删除
 * @return
 */
@ResponseBody
@ApiOperation("员工删除")
@RequestMapping(value = "/employee/delete/{id}",method = RequestMethod.DELETE)
public CommonResult employeeDelete(@PathVariable Integer id){
    employeeService.deleteById(id);
    CommonResult commonResult = CommonResult.success();
    return commonResult;
}
```

这里将id作为路径参数,使用@PathVariable注解注明,删除信息使用Delete方法。

2.4.6 工资管理

(1)工资查询

工资查询功能的代码如下:

```java
/**
 * 工资查询
 * @return
 */
@ResponseBody
@ApiOperation("工资查询")
@RequestMapping(value = "/salary/list",method = RequestMethod.POST)
public PageCommonResult workList(@RequestBody SalaryQO salaryQO){
    PageHelper.startPage(salaryQO.getPage(),salaryQO.getLimit()); // 设置分页参数
    List<SalaryVO> salaryVOS = salaryService.list(salaryQO);       // 查询
    PageInfo<SalaryVO> pageInfo = new PageInfo(salaryVOS);
    //返回结果
    PageCommonResult commonResult = PageCommonResult.success(pageInfo.getList(),pageInfo.getTotal());
    return commonResult;
}
```

这里的分页查询和之前的分页查询基本类似。

（2）生成工资或修改工资

生成工资或修改工资的代码如下：

```java
/**
 * 生成或修改工资
 * @return
 */
@ResponseBody
@ApiOperation("生成或修改工资")
@RequestMapping(value = "/salary/generate",method = RequestMethod.POST)
public CommonResult imitateCheck(@RequestBody Salary salary){
    if(salary.getId() != null &&  !salary.getId().equals("")){
        salaryService.edit(salary);              //修改
    }else{
        salaryService.add(salary);               //新增
    }
    CommonResult commonResult = CommonResult.success();
    return commonResult;
}
```

2.4.7 考勤记录管理

（1）考勤查询

考勤查询的代码如下：

```java
/**
 * 考勤查询
 * @return
 */
@ResponseBody
@ApiOperation("考勤查询")
@RequestMapping(value = "/worktime/list",method = RequestMethod.POST)
```

```java
public PageCommonResult workList(@RequestBody WorktimeQO worktime){
    PageHelper.startPage(worktime.getPage(),worktime.getLimit());
    if(worktime.getCheckTime()!=null){
        Date checkTime = worktime.getCheckTime();
        Calendar checkCalender = Calendar.getInstance();
        checkCalender.setTime(checkTime);
        checkCalender.set(Calendar.HOUR,0);
        checkCalender.set(Calendar.MINUTE,0);
        checkCalender.set(Calendar.SECOND,0);
        Date startTime = checkCalender.getTime();
        checkCalender.add(Calendar.DAY_OF_YEAR,1);
        Date endTime = checkCalender.getTime();
        worktime.setCheckStartTime(startTime);
        worktime.setCheckEndTime(endTime);
    }
    List<WorktimeVO> worktimeVOList = worktimeService.list(worktime);
    PageInfo<WorktimeVO> pageInfo = new PageInfo(worktimeVOList);
    //返回结果
    PageCommonResult commonResult = PageCommonResult.success(pageInfo.getList(),pageInfo.getTotal());
    return commonResult;
}
```

查询时需要用Calendar类转换查询时间。这种时间转换在实际项目中经常会遇到。时间参数设置完成后,进行正常的分页查询即可。不同于前面的是,这里使用了时间区间查询,涉及大于和小于号,与XML中的括号冲突,所以在SQL语句中使用了<![CDATA[...]]>将内容标记为纯文本。

```xml
<select id="list" parameterType="com.lhz.hrms.qo.WorktimeQO" resultMap="VOResultMap">
    select
    wt.id id,wt.check_time,wt.status,wt.check_type,wt.employee_id e_id,em.em_name
    from worktime wt
    left join employee em on wt.employee_id = em.id
    <where>
        <if test="checkStartTime != null">
            <![CDATA[ and wt.check_time >= #{checkStartTime ,jdbcType=TIMESTAMP} ]]>
        </if>
        <if test="checkEndTime != null">
            <![CDATA[ and wt.check_time < #{checkEndTime ,jdbcType=TIMESTAMP} ]]>
        </if>
        <if test="dept != null">
            and em.dept = #{dept ,jdbcType=INTEGER}
        </if>
        <if test="employeeId != null">
            and wt.employee_id = #{employeeId ,jdbcType=INTEGER}
        </if>
    </where>
</select>
```

<![CDATA[...]]> 是XML语法。所有XML文档中的文本均会被解析器解析。只有 CDATA 区段（CDATA section）中的文本会被解析器忽略。如果文本包含了很多的"<"字符"<="和"&"字符，与文本标记语言中的符号冲突，就可以使用CDATA来躲避解析。常见的应用场景有MyBatis中的mapper文件、前端页面脚本文件等。在前端页面的脚本文件中使用，可以采用如下形式：

```
<script>
<![CDATA[
    ...code...
]]>
</script>
```

（2）模拟考勤

模拟考勤的代码如下：

```
/**
 * 模拟考勤
 * @return
 */
@ResponseBody
@ApiOperation("模拟考勤")
@RequestMapping(value = "/worktime/imitateCheck",method = RequestMethod.POST)
public CommonResult imitateCheck(@RequestBody Worktime worktime){
    worktimeService.imitateCheck(worktime);
    CommonResult commonResult = CommonResult.success();
    return commonResult;
}
```

模拟考勤信息并存入数据库。

（3）修改考勤

修改考勤的代码如下：

```
/**
 * 修改考勤为正常状态
 * @return
 */
@ResponseBody
@ApiOperation("修改考勤为正常状态")
@RequestMapping(value = "/work/changeStatus/{workId}",method = RequestMethod.PUT)
public CommonResult imitateCheck(@PathVariable Integer workId){
    worktimeService.changeStatus(workId);
    CommonResult commonResult = CommonResult.success();
    return commonResult;
}
```

2.5 项目总结

  本章主要通过一个较简单的员工管理系统项目向读者介绍如何使用开发工具IntelliJ IDEA搭建Spring Boot开发项目，集成开发所需的前后端框架。通过这样一个简单的项目来了解使用Spring Boot进行项目开发的流程以及一些注意事项。读者在经过本章的学习后，应当亲手编写项目并多加练习。

  本章的重点技术：

- Spring MVC框架：四种请求方式Get、Put、Post、Delete的实现方式和应用场景。
- MyBatis框架：一个持久层框架，如何在Spring Boot中使用MyBatis以及它的作用原理。
- PageHelper插件：如何通过MyBatis引入PageHelper插件以及PageHelper的分页原理。
- LayUI框架：前端页面展示框架。
- 异步加载数据：通过Ajax实现异步获取数据并加载到页面中。
- MD5算法：用在用户密码加密上，防止信息泄露。

第 3 章

二手房管理系统

在第2章中，我们介绍了搭建Spring Boot开发环境的方法，使用Spring、Spring MVC、MyBatis框架组合进行了简单的项目开发。本章在前一章的基础上，开始进行复杂项目的练习。本章涉及的项目是二手房管理系统，后端使用的技术主要有Spring MVC、Spring Data JPA等，前端使用的JS框架为Node.js，UI框架为基于Vue（Vue.js的简称）的Element UI。

本章主要涉及的知识点有：

- 如何使用Spring Boot集成Spring Data JPA。
- 如何使用Spring Data JPA完成数据库操作。
- 如何使用Node.js前端框架进行数据交互。

3.1 项目技术选型

从本章开始，所涉及的架构开始变得复杂。本章涉及的二手房管理系统，后端选用的框架是Spring MVC、Spring Data JPA，数据库使用MySQL，前端选用的框架是Node.js和Element UI。

3.1.1 Spring Data JPA

JPA全称是Java Persistence API，是Java针对持久化指定的数据访问规范。Spring Data JPA是基于该规范实现的数据访问层框架。使用Spring Data JPA可以简化数据库操作，开发者只需要编写数据访问接口，Spring Data JPA会自动进行实现。读者可到Spring Data JPA官网查阅相关的介绍文档和API使用文档，也可以下载其源码进行学习。官方网址为https://spring.io/projects/spring-data-jpa。

使用Spring Data JPA需要JDK8+以上的版本。如果使用Maven构建项目，那么在pom文件中引入如下依赖，然后在Spring Boot启动类上加上@EnableJpaAuditing注解即可。

```
<dependency>
    <groupId>org.springframework.boot</groupId>
    <artifactId>spring-boot-starter-data-jpa</artifactId>
</dependency>
@SpringBootApplication
@EnableJpaAuditing
public class HouseApplication {
    public static void main(String[] args) {
        SpringApplication.run(HouseApplication.class, args);
    }
}
```

Spring Data JPA为我们提供Repository层的接口，支持两种方式，一种是基于方法名称查询，另一种是基于@Query注解查询。

1. 基于方法名称查询

在自定义的Repository接口中继承JpaRepository，然后按照规范编写对应的方法名，Spring Data JPA会自动转化为对应的SQL语句并执行数据库操作，封装返回结果。

常用的命名规范如表3.1所示。

表3.1　Spring Data JPA常用命名规范

关键字	方法名举例	对应SQL语句
Distinct	findDistinctByLastnameAndFirstname	select distinct … where x.lastname = ?1 and x.firstname = ?2
And	findByLastnameAndFirstname	… where x.lastname = ?1 and x.firstname = ?2
Or	findByLastnameOrFirstname	… where x.lastname = ?1 or x.firstname = ?2
Is, Equals	findByFirstname,findByFirstnameIs, findByFirstnameEquals	… where x.firstname = ?1
Between	findByStartDateBetween	… where x.startDate between ?1 and ?2
LessThan	findByAgeLessThan	… where x.age < ?1
LessThanEqual	findByAgeLessThanEqual	… where x.age <= ?1
GreaterThan	findByAgeGreaterThan	… where x.age > ?1
GreaterThanEqual	findByAgeGreaterThanEqual	… where x.age >= ?1
After	findByStartDateAfter	… where x.startDate > ?1
Before	findByStartDateBefore	… where x.startDate < ?1
IsNull, Null	findByAge(Is)Null	… where x.age is null
IsNotNull, NotNull	findByAge(Is)NotNull	… where x.age not null
Like	findByFirstnameLike	… where x.firstname like ?1
NotLike	findByFirstnameNotLike	… where x.firstname not like ?1
StartingWith	findByFirstnameStartingWith	… where x.firstname like ?1 (parameter bound with appended %)
EndingWith	findByFirstnameEndingWith	… where x.firstname like ?1 (parameter bound with prepended %)

(续表)

关 键 字	方法名举例	对应SQL语句
Containing	findByFirstnameContaining	… where x.firstname like ?1 (parameter bound wrapped in %)
OrderBy	findByAgeOrderByLastnameDesc	… where x.age = ?1 order by x.lastname desc
Not	findByLastnameNot	… where x.lastname <> ?1
In	findByAgeIn(Collection<Age> ages)	… where x.age in ?1
NotIn	findByAgeNotIn(Collection<Age> ages)	… where x.age not in ?1
True	findByActiveTrue()	… where x.active = true
False	findByActiveFalse()	… where x.active = false
IgnoreCase	findByFirstnameIgnoreCase	… where UPPER(x.firstname) = UPPER(?1)

除了常用的关键字findBy以外，Spring Data JPA还支持一些其他关键字，如表3.2所示。

表3.2　Spring Data JPA方法名常用关键字

关 键 字	描　　述
find…By, read…By, get…By, query…By, search…By, stream…By	查询方法，例如 findBy…, findMyDomainTypeBy…，也可以和其他关键字组合，例如findFirstById
exists…By	判断数据是否存在，返回布尔型结果
count…By	返回结果集的size
delete…By, remove…By	删除方法
…First<number>…, …Top<number>…	根据关键字返回结果集的子集或某一个对象
…Distinct…	根据关键字对结果集进行去重

2．使用@Query注解查询

除了基于方法名称的查询之外，还可以使用@Query注解传递SQL语句，用来实现比较复杂的场景。

例如一个逻辑删除方法，通过@Query传递SQL语句，使用占位符设置参数：

```
@Modifying
@Transactional
@Query(value = "update service_banner_info set del_flag =?1 where id =?2",
nativeQuery = true)
void delete(String delFlag, String id);
```

又如：

```
public interface UserRepository extends JpaRepository<User, Long> {
  @Query("select u from User u where u.firstname = :firstname or u.lastname = :lastname")
  User findByLastnameOrFirstname(@Param("lastname") String lastname,
  @Param("firstname") String firstname);
```

这里使用命名参数。使用@Param注解声明参数名，在SQL语句中使用冒号加参数名的形式赋值。

Spring Data JPA支持JPA规范中定义的注解。常用的注解可以参考表3.3。

表3.3　Spring Data JPA常用注解

注 解 名	作　　用
@Entity	标识类为数据库持久化对象实体
@Table	标识实体类在数据库所对应的表名
@Id	标识变量为主键
@GeneratedValue	设置主键生成策略
@Column	标识属性所对应的字段名，可进行自定义
@Transient	标识属性并非数据库表字段的映射
@Temporal	时间日期类型，用这个注释来注明具体的数据库类型，TemporalType.DATE、TemporalType.TIME、TemporalType.TIMESTAMP
@Enumerated	使用此注解映射枚举字段，EnumType.ORDINAL、EnumType.STRING
@Embeddable @Embedded	当一个实体类要在多个不同的实体类中使用而不需要生成数据库表时： ● @Embeddable：注解在类上，表示此类可以被其他类嵌套。 ● @Embedded：注解在属性上，表示嵌套被@Embeddable注解的同类型类
@ElementCollection	集合映射
@CreatedDate @CreatedBy @LastModifiedDate @LastModifiedBy	表示字段为创建时间字段（insert自动设置）、创建用户字段（insert自动设置）、最后修改时间字段（update自定设置）、最后修改用户字段（update自定设置）

这些注解可以联合使用。例如使用@Query查询时，如果在实体类上添加了@Entity注解，那么@Query注解中可以使用动态变量#{entityName}操作对应数据库表。

```
@Entity public class User {
    @Id
    @GeneratedValue
    Long id;
    String lastname;
}

public interface UserRepository extends JpaRepository<User,Long> {
@Query("select u from #{#entityName} u where u.lastname = ?1")
List<User> findByLastname(String lastname);
}
```

3. 排序

如果想使用排序查询，可在查询方法内传入Sort参数，调用时根据具体业务传入要排序的规则即可。这里同时支持Sort和JpaSort对象。

```
public interface UserRepository extends JpaRepository<User, Long> {

  @Query("select u from User u where u.lastname like ?1%")
  List<User> findByAndSort(String lastname, Sort sort);

  @Query("select u.id, LENGTH(u.firstname) as fn_len from User u where u.lastname like ?1%")
```

```
 List<Object[]> findByAsArrayAndSort(String lastname, Sort sort);
}
repo.findByAndSort("lannister", Sort.by("firstname"));
repo.findByAndSort("stark", Sort.by("LENGTH(firstname)"));
repo.findByAndSort("targaryen", JpaSort.unsafe("LENGTH(firstname)"));
repo.findByAsArrayAndSort("bolton", Sort.by("fn_len"));
```

由于篇幅有限，关于Spring Data JPA的详细使用，读者可参考Spring官方文档。

3.1.2 Node.js

Node.js是一个开源与跨平台的 JavaScript 运行时环境，基于Google的V8引擎，使得它的速度快、性能好。借助Node.js可以单独开启前端服务。

在使用Node.js前需要安装Node.js。具体操作步骤如下：

步骤01 在Node.js的官网找到对应的版本下载，如图3.1所示。

Windows 安装包 (.msi)	32 位	64 位
Windows 二进制文件 (.zip)	32 位	64 位
macOS 安装包 (.pkg)	64 位	
macOS 二进制文件 (.tar.gz)	64 位	
Linux 二进制文件 (x64)	64 位	
Linux 二进制文件 (ARM)	ARMv7	ARMv8
Docker 镜像	官方镜像	
全部安装包	阿里云镜像	

图3.1　Node.js下载页面

笔者使用的操作系统为Windows 10 64位，所以这里以Windows 64位安装包为例安装Node.js。

步骤02 双击安装包（见图3.2）运行安装。所有选项都选择默认项。

| node-v16.6.1-x64.msi | 2021/8/11 15:14 | Windows Install... | 28,052 KB | Installer |

图3.2　Node.js安装包

步骤03 安装完成后会自动弹出配置窗口，如图3.3所示，等待自动配置即可，无须任何操作。
步骤04 配置完成后查看安装目录，可以看到如图3.4所示的内容。
步骤05 使用cmd命令进行简单的验证，分别输入node -v和npm -v，如果能够显示版本信息，则表示安装成功，如图3.5所示。

接下来进行环境变量配置。

步骤06 在Node.js的安装目录下新建两个文件夹，node_global及node_cache。这里的环境配置主要配置的是npm安装的全局模块所在的路径，以及缓存cache的路径。
步骤07 创建完成后进入cmd命令窗口，输入如下两行命令：

```
npm config set prefix "D:\Program Files\nodejs\node_global"
npm config set cache "D:\Program Files\nodejs\node_cache"
```

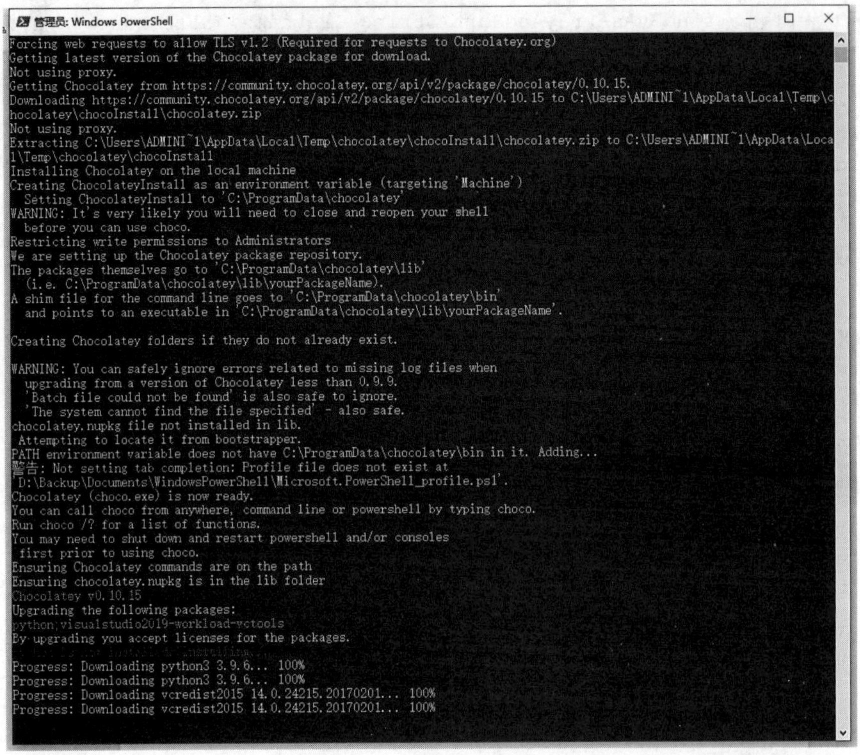

图3.3 Node.js安装页面

图3.4 Node.js安装目录

图3.5 Node.js安装验证

步骤08 打开"编辑系统变量"对话框，新建NODE_PATH变量，变量值为D:\Program Files\nodejs\node_global\node_modules（见图3.6），编辑用户变量下的Path变量，将其中默认的npm路径修改为D:\Program Files\nodejs\node_global，如图3.7和图3.8所示。

图3.6 配置Node.js环境变量（1）

图3.7 配置Node.js环境变量（2）

图3.8 配置Node.js环境变量（3）

至此，Node.js就安装并配置完成了。

3.1.3 Vue和Element UI

本章项目前端主要使用的是基于Vue的Element UI框架。Vue是一套用于构建用户页面的渐进式框架，与其他大型框架不同的是，Vue被设计为可以自底向上逐层应用。一方面Vue的核心库只关注视图层，不仅易于上手，还便于与第三方库或既有项目整合。另一方面，当与现代化的工具链以及各种支持类库结合使用时，Vue也完全能够为复杂的单页应用提供驱动。

Element UI是一套基于Vue的组件库，能帮助开发者快速开发项目。Element UI有以下几个特点：

（1）一致性

- 与现实生活一致：与现实生活的流程、逻辑保持一致，遵循用户习惯的语言和概念。
- 在页面中一致：所有的元素和结构需保持一致，比如，设计样式、图标和文本、元素的位置等。

（2）反馈

- 控制反馈：通过页面样式和交互动效让用户可以清晰地感知自己的操作。
- 页面反馈：操作后，通过页面元素的变化清晰地展现当前状态。

（3）高效率

- 简化流程：设计简洁直观的操作流程。
- 清晰明确：语言表达清晰且表意明确，能让用户快速理解进而作出决策。
- 帮助用户识别：页面简单直白，让用户快速识别而非回忆，减少用户记忆负担。

（4）可控性

- 用户决策：根据场景可给予用户操作建议或安全提示，但不能代替用户进行决策。
- 结果可控：用户可以自由地进行操作，包括撤销、回退和终止当前操作等。

使用Element UI只需在main.js中引入响应的组件库即可。示例代码如下：

```
import Vue from 'vue';
import ElementUI from 'element-ui';
import 'element-ui/lib/theme-chalk/index.css'
;import App from './App.vue';
Vue.use(ElementUI);
new Vue({
  el: '#app',
  render: h => h(App)
});
```

由于本书主要讲解Spring Boot开发实战，所以对前端技术不再大幅描述。读者可参考对应的官方文档。

3.1.4 框架搭建

本章项目选取的是二手房管理系统，以下为该项目的技术框架部分：

- 开发框架：Spring Boot。
- 数据库：MySQL。
- 后台框架：Spring、Spring MVC、Spring Data JPA。
- 前端框架：Element UI。
- 前端JS框架：Node.js。

本项目分为两部分，后端服务项目和前端页面项目。因为本书主要是讲解Spring Boot的使用，所以着重介绍后端服务框架的搭建过程，不再详细描述前端框架的搭建过程。对Node.js感兴趣的读者可以深入研究配套源码，或自行搜寻相关资料进行学习。

创建项目部分可参考第2章。创建完成后，引入相关依赖。除了第2章所涉及的部分依赖外，这里需要引入Spring Data JPA的依赖。创建完成后，主要的pom文件如下：

```
<dependency>
    <groupId>org.springframework.boot</groupId>
    <artifactId>spring-boot-starter</artifactId>
</dependency>
<!--Spring Data JPA-->
<dependency>
```

```xml
    <groupId>org.springframework.boot</groupId>
    <artifactId>spring-boot-starter-data-jpa</artifactId>
</dependency>
<dependency>
    <groupId>org.springframework.boot</groupId>
    <artifactId>spring-boot-starter-web</artifactId>
</dependency>
<dependency>
    <groupId>mysql</groupId>
    <artifactId>mysql-connector-java</artifactId>
    <version>8.0.22</version>
</dependency>
<!--JSON支持-->
<dependency>
    <groupId>com.alibaba</groupId>
    <artifactId>fastjson</artifactId>
    <version>1.2.9</version>
</dependency>
<dependency>
    <groupId>net.minidev</groupId>
    <artifactId>json-smart</artifactId>
    <version>2.3</version>
    <scope>compile</scope>
</dependency>
<dependency>
    <groupId>net.sf.json-lib</groupId>
    <artifactId>json-lib</artifactId>
    <version>2.4</version>
    <classifier>jdk15</classifier>
</dependency>
<!--lombok插件-->
<dependency>
    <groupId>org.projectlombok</groupId>
    <artifactId>lombok</artifactId>
    <version>1.16.16</version>
</dependency>
```

接下来配置Spring Boot的配置文件。上一章我们采用的是application.properties形式的配置文件。本章我们采用yml形式的配置文件。相对于properties文件，yml文件更具层次感，是一种树形结构，可读性更强。yml文件的主要内容如下：

```yml
spring:
  http:
  datasource:
    url: jdbc:mysql://localhost:3306/ershoufang?characterEncoding=utf8&useSSL=false&serverTimezone=Asia/Shanghai&rewriteBatchedStatements=true   #数据库地址
    username: root   #用户名
    password: 123456   #密码
    hikari:
      driver-class-name: com.mysql.cj.jdbc.Driver   #驱动
  mvc:
```

```yaml
      pathmatch:
        matching-strategy: ant_path_matcher
  jpa:
    properties:
      hibernate:
        hbm2ddl:
          auto: update              #启动时自动更新表结构
      show-sql: true                #显示SQL语句
  servlet:
    multipart:
      enabled: true                 #支持多媒体
      max-file-size: 20000MB        #文件大小上限
      max-request-size: 20000MB     #请求大小上限
  jooq:
    sql-dialect: org.hibernate.dialect.MySQL5Dialect    #SQL方言
server:
  servlet:
    session:
      timeout: 30m                  #超时时间
  port: 8176                        #端口
```

注意：变量后写值时，中间一定要有一个空格。例如，端口port，值为8176，那么8176与前面冒号中间一定要有空格。

由于本项目涉及图片上传，所以接下来在配置文件中配置文件上传相关的信息。在resources目录下创建config.properties文件，文件内容如下：

```
##windows
##文件存放目录
filePath=E:/upload/ershoufang/
//filePath=E:/upload/ershoufang/
##文件上传安全域, 前端项目地址
fileCROS=http://localhost:8067
```

配置完成后需要编写一个类来获取这些变量，该类放置在config目录下。

```java
package com.xsl.service.config;
import lombok.Data;
import org.springframework.beans.factory.annotation.Value;
import org.springframework.context.annotation.PropertySource;
import org.springframework.stereotype.Component;
@Data
@Component
@PropertySource({"classpath:config.properties"})
public class ConfigProperties {
    @Value("${filePath}")
    private String filePath;              //文件存放目录
    @Value("${fileCROS}")
    private String fileCros;              //文件上传安全域, 前端项目地址
    @Value("${userDemoFile}")
    private String userDemoFile;          //用户示例文件
    @Value("${templateFile}")
```

```
    private String templateFile;              //模板文件
    @Value("${demoPath}")
    private String demoPath;                   //示例路径
    @Value("${applyInfoFile}")
    private String applyInfoFile;              //申请信息
    @Value("${imgUrl}")
    private String imgUrl;                     //图片路径
    @Value("${upload}")
    private String upload;                     //上传地址
    @Value("${accessKey}")
    private String accessKey;                  //通行key
    @Value("${accessKeySecret}")
    private String accessKeySecret;            //通行秘钥
    @Value("${adminList}")
    private String adminList;                  //管理员列表
    @Value("${SMSId}")
    private String SMSId;                      //smsId
}
```

另外创建所需的各个目录待开发时使用。后端框架搭建完成后的基本结构如图3.9所示。前端项目使用Element UI+Node.js的架构，代码结构如图3.10所示。

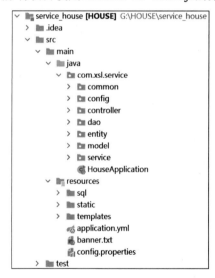
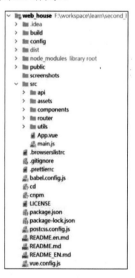

图3.9　二手房管理系统后端代码结构　　　　图3.10　二手房管理系统前端代码结构

至此，基本开发环境已搭建完成，接下来进入系统开发阶段。

3.2　项目前期准备

3.2.1　项目需求说明

本章涉及的二手房管理系统，主要包含的功能有二手房房源管理、房源信息管理、楼盘

信息管理、楼盘动态管理、认购管理、销售管理、认筹管理、楼盘收藏、个人中心、系统管理、配置管理等。能够实现二手房信息管理、楼盘信息同步,对用户操作进行分析,例如用户收藏的楼盘等,精准掌握用户偏好,促成交易。还能实现对二手房交易做记录,包括交易信息、合同信息等。

3.2.2 系统功能设计

根据二手房管理系统要实现的主要功能,设计对应的功能列表。该系统的系统功能结构如图3.11所示。

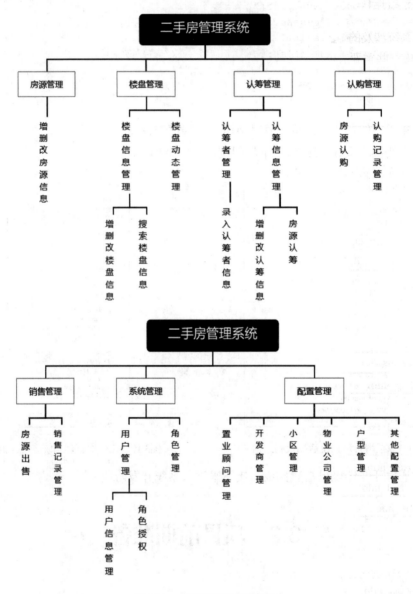

图3.11 二手房管理系统结构

- **房源管理**:实现二手房房源信息的增、删、改、查功能。

- 楼盘管理：实现楼盘信息的增、删、改、查功能，并及时更新楼盘动态。
- 认筹管理：认筹信息的增、删、改、查；认筹者的增、删、改、查功能。
- 认购管理：录入用户认购信息，修改认购信息。
- 销售管理：录入出售房源记录，修改、删除、查询销售信息，维护合同附件信息。
- 系统管理：包括用户管理和角色管理。
- 配置管理：包括置业顾问管理、开发商管理、小区管理、物业公司管理、户型管理、装修类型管理、建筑类型管理等。

3.2.3 系统数据库设计

由于本系统涉及的数据库数量比较多，总计达38个，如表3.4所示，这里只简单介绍每个表的功能，并列出一些重要表的建表语句，剩余表的建表语句请读者参考随书配套的相关SQL文件。

表3.4 二手房管理系统数据库表

表 名	含 义
sys_user	系统用户表，用于系统登录、权限校验
sys_role	系统角色表，用于用户角色设置
sys_dic	系统字典表，用于维护常见选项
sys_setting_doc	设置文档，包括用户隐私协议、关于我们等
two_hand_house_info	二手房房源基础信息表
two_hand_house_info_facilities	房子配套设施关系表，关联数据字典
two_hand_ordinary_residence	房源对应的普通住宅信息表
service_user_real_estate	用户收藏的楼盘信息
service_user_house	用户房源关联表
service_subscription	认购信息表
service_sale_info	销售信息表
service_banner_info	页面banner显示图片表
service_building_info	楼栋信息表
service_building_tag	建筑标签表
service_building_type	建筑类型表
service_community_info	小区信息表
service_company_info	公司信息表
service_consultant_info	置业顾问表
service_coupon_info	优惠券信息表
service_coupon_info_real_estates	优惠券和楼盘关联表
service_coupon_info_users	优惠券和用户关联表
service_decoration_info	装修信息表
service_developer_info	开发商信息表
service_fund_raising	认筹信息表
service_fund_raising_user	用户认筹关联表

(续表)

表 名	含 义
service_help_center_info	帮助信息表
service_house_info	房源信息表
service_house_type_info	房源类型表
service_property_info	物业信息表
service_property_type	物业类型表
service_real_estate_dynamic_info	楼盘动态表
service_real_estate_info	楼盘信息表
service_real_estate_info_building_tags	楼盘建筑标签关联表
service_real_estate_info_building_types	楼盘建筑类型关联表
service_real_estate_info_consultants	楼盘置业顾问关联表
service_real_estate_info_decoration_infos	楼盘装修信息表
service_real_estate_info_house_type_infos	楼盘房源类型表
service_real_estate_info_property_types	楼盘物业关联表

（1）系统用户表，表名sys_user，用于登录。

```
CREATE TABLE 'sys_user' (
  'id' varchar(32) CHARACTER SET utf8mb4 COLLATE utf8mb4_unicode_ci NOT NULL,
  'create_date' datetime(6) DEFAULT NULL,
  'del_flag' int NOT NULL,
  'update_date' datetime(6) DEFAULT NULL,
  'name' varchar(255) CHARACTER SET utf8mb4 COLLATE utf8mb4_unicode_ci DEFAULT NULL,
  'password' varchar(255) CHARACTER SET utf8mb4 COLLATE utf8mb4_unicode_ci DEFAULT NULL,
  'phone' varchar(255) CHARACTER SET utf8mb4 COLLATE utf8mb4_unicode_ci DEFAULT NULL,
  'remark' varchar(255) CHARACTER SET utf8mb4 COLLATE utf8mb4_unicode_ci DEFAULT NULL,
  'role_type' varchar(255) CHARACTER SET utf8mb4 COLLATE utf8mb4_unicode_ci DEFAULT NULL,
  'sort' varchar(255) CHARACTER SET utf8mb4 COLLATE utf8mb4_unicode_ci DEFAULT NULL,
  'user_name' varchar(255) CHARACTER SET utf8mb4 COLLATE utf8mb4_unicode_ci DEFAULT NULL,
  'sex' varchar(255) CHARACTER SET utf8mb4 COLLATE utf8mb4_unicode_ci DEFAULT NULL,
  'verification_code' varchar(255) CHARACTER SET utf8mb4 COLLATE utf8mb4_unicode_ci DEFAULT NULL,
  'login_time' datetime(6) DEFAULT NULL,
  'verification_time' datetime(6) DEFAULT NULL,
  'verification_code_forget' varchar(255) CHARACTER SET utf8mb4 COLLATE utf8mb4_unicode_ci DEFAULT NULL,
  'verification_time_forget' datetime(6) DEFAULT NULL,
  'company_id' varchar(255) CHARACTER SET utf8mb4 COLLATE utf8mb4_unicode_ci DEFAULT NULL,
  'com_manager' int NOT NULL DEFAULT '0' COMMENT '公司管理员',
```

```
    'id_card' varchar(255) CHARACTER SET utf8mb4 COLLATE utf8mb4_unicode_ci DEFAULT NULL,
    'buy_date' datetime(6) DEFAULT NULL,
    'buy_money' varchar(255) CHARACTER SET utf8mb4 COLLATE utf8mb4_unicode_ci DEFAULT NULL,
    'offer_date' datetime(6) DEFAULT NULL,
    'offer_money' varchar(255) CHARACTER SET utf8mb4 COLLATE utf8mb4_unicode_ci DEFAULT NULL,
    'headimgurl' varchar(255) CHARACTER SET utf8mb4 COLLATE utf8mb4_unicode_ci DEFAULT NULL,
    'nickname' varchar(255) CHARACTER SET utf8mb4 COLLATE utf8mb4_unicode_ci DEFAULT NULL,
    'openid' varchar(255) CHARACTER SET utf8mb4 COLLATE utf8mb4_unicode_ci DEFAULT NULL,
    PRIMARY KEY ('id') USING BTREE
) ENGINE=InnoDB DEFAULT CHARSET=utf8mb4 COLLATE=utf8mb4_unicode_ci ROW_FORMAT=DYNAMIC;
```

（2）数据字典表，表名sys_dice，用于维护一些基础配置，例如配套设施、房产年限、装修程度、户型等。

```
CREATE TABLE 'sys_dic' (
    'id' varchar(32) CHARACTER SET utf8mb4 COLLATE utf8mb4_general_ci NOT NULL,
    'create_date' datetime DEFAULT NULL,
    'del_flag' int NOT NULL,
    'update_date' datetime DEFAULT NULL,
    'level' varchar(255) CHARACTER SET utf8mb4 COLLATE utf8mb4_general_ci DEFAULT NULL,
    'parent_id' varchar(255) CHARACTER SET utf8mb4 COLLATE utf8mb4_general_ci DEFAULT NULL,
    'sort' int DEFAULT NULL,
    'title' varchar(255) CHARACTER SET utf8mb4 COLLATE utf8mb4_general_ci DEFAULT NULL,
    PRIMARY KEY ('id') USING BTREE
) ENGINE=InnoDB DEFAULT CHARSET=utf8mb4 COLLATE=utf8mb4_general_ci ROW_FORMAT=DYNAMIC;
```

（3）二手房房源表，表名two_hand_house_info。

```
CREATE TABLE 'two_hand_house_info' (
    'id' varchar(32) CHARACTER SET utf8mb4 COLLATE utf8mb4_general_ci NOT NULL,
    'create_date' datetime DEFAULT NULL,
    'del_flag' int NOT NULL,
    'update_date' datetime DEFAULT NULL,
    'community_id' varchar(32) CHARACTER SET utf8mb4 COLLATE utf8mb4_unicode_ci DEFAULT NULL,
    'actual_area' varchar(100) CHARACTER SET utf8mb4 COLLATE utf8mb4_general_ci DEFAULT '' COMMENT '实际面积',
    'address' varchar(512) CHARACTER SET utf8mb4 COLLATE utf8mb4_general_ci DEFAULT '' COMMENT '地址',
    'age_type' int DEFAULT '0' COMMENT '房源类型：0 未满2年;1 满两年;2 满5年',
    'agency_fee_percentage' varchar(8) CHARACTER SET utf8mb4 COLLATE utf8mb4_general_ci DEFAULT ' ' COMMENT ' 佣金比例 ',
    'certificate_area' varchar(100) CHARACTER SET utf8mb4 COLLATE utf8mb4_general_ci DEFAULT '' COMMENT '产证面积',
```

```
    'community_img_urls' varchar(1000) CHARACTER SET utf8mb4 COLLATE
utf8mb4_general_ci DEFAULT '' COMMENT '小区图',
    'community_matching' varchar(1000) CHARACTER SET utf8mb4 COLLATE
utf8mb4_general_ci DEFAULT '' COMMENT '小区配套',
    'content' varchar(1000) CHARACTER SET utf8mb4 COLLATE utf8mb4_general_ci DEFAULT
'' COMMENT '房源描述',
    'custom_no' varchar(100) CHARACTER SET utf8mb4 COLLATE utf8mb4_general_ci
DEFAULT '' COMMENT '自定义编号',
    'direction' varchar(10) CHARACTER SET utf8mb4 COLLATE utf8mb4_general_ci DEFAULT
'' COMMENT '朝向',
    'floor_belong' int DEFAULT '0' COMMENT '楼层 所属',
    'floor_total' int DEFAULT '0' COMMENT '楼层 总数',
    'house_age' int DEFAULT '0' COMMENT '房龄',
    'house_no' varchar(200) CHARACTER SET utf8mb4 COLLATE utf8mb4_general_ci DEFAULT
'' COMMENT '房源编号',
    'house_type_balcony' int DEFAULT '0' COMMENT '房型 阳台',
    'house_type_hall' int DEFAULT '0' COMMENT '房型 厅',
    'house_type_img_urls' varchar(1000) CHARACTER SET utf8mb4 COLLATE
utf8mb4_general_ci DEFAULT '' COMMENT '户型图',
    'house_type_kitchen' int DEFAULT '0' COMMENT '房型 厨房',
    'house_type_room' int DEFAULT '0' COMMENT '房型 室',
    'house_type_toilet' int DEFAULT '0' COMMENT '房型 卫',
    'inside_img_urls' varchar(1000) CHARACTER SET utf8mb4 COLLATE utf8mb4_general_ci
DEFAULT '' COMMENT '室内图',
    'is_elevator' int DEFAULT '0' COMMENT '电梯 1 是;2 否',
    'is_foreign_style' int DEFAULT '0' COMMENT '洋房 1 是;2 不是',
    'is_key' int DEFAULT '0' COMMENT '有钥匙 1 有;2 没有',
    'is_new' int DEFAULT '0' COMMENT '新房 1 是;2 不是',
    'is_only' int DEFAULT '0' COMMENT '唯一类型: 0 不唯一 ;1 唯一住房',
    'key_num' varchar(100) CHARACTER SET utf8mb4 COLLATE utf8mb4_general_ci DEFAULT
'' COMMENT '钥匙编号',
    'no_agency_fee' int DEFAULT '0' COMMENT '不收中介费 1 是;2 否',
    'owner_address' varchar(100) CHARACTER SET utf8mb4 COLLATE utf8mb4_general_ci
DEFAULT '' COMMENT '房东 地址',
    'owner_build' varchar(100) CHARACTER SET utf8mb4 COLLATE utf8mb4_general_ci
DEFAULT '' COMMENT '房东 楼号',
    'owner_community' varchar(100) CHARACTER SET utf8mb4 COLLATE utf8mb4_general_ci
DEFAULT '' COMMENT '房东 单元',
    'owner_house' varchar(20) CHARACTER SET utf8mb4 COLLATE utf8mb4_general_ci
DEFAULT '' COMMENT '房东 房号',
    'owner_mentality' varchar(1000) CHARACTER SET utf8mb4 COLLATE utf8mb4_general_ci
DEFAULT '' COMMENT '业主心态',
    'owner_name' varchar(100) CHARACTER SET utf8mb4 COLLATE utf8mb4_general_ci
DEFAULT '' COMMENT '房东 联系人',
    'owner_remark' varchar(1000) CHARACTER SET utf8mb4 COLLATE utf8mb4_general_ci
DEFAULT '' COMMENT '房东 备注信息',
    'owner_tel' varchar(100) CHARACTER SET utf8mb4 COLLATE utf8mb4_general_ci
DEFAULT '' COMMENT '房东 电话',
    'sale_price' varchar(100) CHARACTER SET utf8mb4 COLLATE utf8mb4_general_ci
DEFAULT '' COMMENT '售价 单位:万元',
    'service_introduction' varchar(1000) CHARACTER SET utf8mb4 COLLATE
utf8mb4_general_ci DEFAULT '' COMMENT '服务介绍',
    'tax_information' varchar(1000) CHARACTER SET utf8mb4 COLLATE utf8mb4_general_ci
DEFAULT '' COMMENT '税费信息',
    'title' varchar(200) CHARACTER SET utf8mb4 COLLATE utf8mb4_general_ci DEFAULT
'' COMMENT '房源标题',
```

```
    'video_img_url' varchar(100) CHARACTER SET utf8mb4 COLLATE utf8mb4_general_ci
DEFAULT ' ' COMMENT '视频图片',
    'video_url' varchar(100) CHARACTER SET utf8mb4 COLLATE utf8mb4_general_ci
DEFAULT ' ' COMMENT '视频地址',
    'place_id' varchar(32) CHARACTER SET utf8mb4 COLLATE utf8mb4_unicode_ci DEFAULT
NULL,
    'ordinary_residence_id' varchar(32) CHARACTER SET utf8mb4 COLLATE
utf8mb4_general_ci DEFAULT NULL,
    PRIMARY KEY ('id') USING BTREE,
    KEY 'FKmyoixa6foukar7b0fcc2r44ax' ('community_id') USING BTREE,
    KEY 'FKlg1edo7ljwx67p9knehowv76v' ('place_id') USING BTREE,
    KEY 'FKnl5gvn25vj2ivrj8at0515b32' ('ordinary_residence_id') USING BTREE,
    CONSTRAINT 'FKlg1edo7ljwx67p9knehowv76v' FOREIGN KEY ('place_id') REFERENCES
'sys_place_info' ('id') ON DELETE RESTRICT ON UPDATE RESTRICT,
    CONSTRAINT 'FKmyoixa6foukar7b0fcc2r44ax' FOREIGN KEY ('community_id')
REFERENCES 'service_community_info' ('id') ON DELETE RESTRICT ON UPDATE RESTRICT,
    CONSTRAINT 'FKnl5gvn25vj2ivrj8at0515b32' FOREIGN KEY ('ordinary_residence_id')
REFERENCES 'two_hand_ordinary_residence' ('id') ON DELETE RESTRICT ON UPDATE RESTRICT
) ENGINE=InnoDB DEFAULT CHARSET=utf8mb4 COLLATE=utf8mb4_general_ci
ROW_FORMAT=DYNAMIC;
```

（4）楼盘信息表，表名service_real_estate_info，用来管理楼盘信息。

```
CREATE TABLE 'service_real_estate_info' (
    'id' varchar(32) CHARACTER SET utf8mb4 COLLATE utf8mb4_unicode_ci NOT NULL,
    'create_date' datetime(6) DEFAULT NULL,
    'del_flag' int NOT NULL,
    'update_date' datetime(6) DEFAULT NULL,
    'address' varchar(255) CHARACTER SET utf8mb4 COLLATE utf8mb4_unicode_ci DEFAULT
'' COMMENT '楼盘地址（省/市/区、详细地址）',
    'brief' text CHARACTER SET utf8mb4 COLLATE utf8mb4_unicode_ci COMMENT '楼盘描
述',
    'feature' text CHARACTER SET utf8mb4 COLLATE utf8mb4_unicode_ci COMMENT '楼盘
特色（纯文本描述）',
    'img_url1' varchar(200) CHARACTER SET utf8mb4 COLLATE utf8mb4_unicode_ci DEFAULT
'' COMMENT '配套 链接',
    'img_url2' varchar(200) CHARACTER SET utf8mb4 COLLATE utf8mb4_unicode_ci DEFAULT
'' COMMENT '区位 链接',
    'img_url3' varchar(200) CHARACTER SET utf8mb4 COLLATE utf8mb4_unicode_ci DEFAULT
'' COMMENT '项目现场 链接',
    'img_url4' varchar(200) CHARACTER SET utf8mb4 COLLATE utf8mb4_unicode_ci DEFAULT
'' COMMENT '样板间 链接',
    'img_url5' varchar(200) CHARACTER SET utf8mb4 COLLATE utf8mb4_unicode_ci DEFAULT
'' COMMENT '预售许可证 链接',
    'img_url6' varchar(200) CHARACTER SET utf8mb4 COLLATE utf8mb4_unicode_ci DEFAULT
'' COMMENT '总规图 链接',
    'open_date' varchar(255) CHARACTER SET utf8mb4 COLLATE utf8mb4_unicode_ci
DEFAULT NULL COMMENT '开盘时间',
    'plan_delivery_date' varchar(255) CHARACTER SET utf8mb4 COLLATE
utf8mb4_unicode_ci DEFAULT '' COMMENT '预计交房时间',
    'plan_offer_date' varchar(255) CHARACTER SET utf8mb4 COLLATE utf8mb4_unicode_ci
DEFAULT '' COMMENT '预计认筹时间',
    'property_right_year' int DEFAULT '0' COMMENT '产权年限',
    'public_qr_code' varchar(255) CHARACTER SET utf8mb4 COLLATE utf8mb4_unicode_ci
DEFAULT '' COMMENT '公众号二维码',
```

```sql
      'surrounding' varchar(2000) CHARACTER SET utf8mb4 COLLATE utf8mb4_unicode_ci 
DEFAULT '' COMMENT '周边配套',
      'title' varchar(255) CHARACTER SET utf8mb4 COLLATE utf8mb4_unicode_ci DEFAULT 
'' COMMENT '楼盘名称',
      'video_url' varchar(255) CHARACTER SET utf8mb4 COLLATE utf8mb4_unicode_ci 
DEFAULT '' COMMENT '视频',
      'we_chat' varchar(255) CHARACTER SET utf8mb4 COLLATE utf8mb4_unicode_ci DEFAULT 
'' COMMENT '微信号',
      'developer_info_id' varchar(32) CHARACTER SET utf8mb4 COLLATE utf8mb4_unicode_ci 
DEFAULT NULL,
      'community_id' varchar(32) CHARACTER SET utf8mb4 COLLATE utf8mb4_unicode_ci 
DEFAULT NULL,
      'building_infos' varchar(255) CHARACTER SET utf8mb4 COLLATE utf8mb4_unicode_ci 
DEFAULT NULL,
      'vr_img_url' varchar(200) CHARACTER SET utf8mb4 COLLATE utf8mb4_unicode_ci 
DEFAULT '' COMMENT 'VR图片链接',
      'vr_url' varchar(200) CHARACTER SET utf8mb4 COLLATE utf8mb4_unicode_ci DEFAULT 
'' COMMENT 'VR链接',
      'average_price' varchar(255) CHARACTER SET utf8mb4 COLLATE utf8mb4_unicode_ci 
DEFAULT '' COMMENT '参考均价',
      'floor_area_range' varchar(255) CHARACTER SET utf8mb4 COLLATE utf8mb4_unicode_ci 
DEFAULT '' COMMENT '建面范围',
      'property_info_id' varchar(32) CHARACTER SET utf8mb4 COLLATE utf8mb4_unicode_ci 
DEFAULT NULL,
      'video_img_url' varchar(200) CHARACTER SET utf8mb4 COLLATE utf8mb4_unicode_ci 
DEFAULT '' COMMENT '视频图片',
      'preferential' text CHARACTER SET utf8mb4 COLLATE utf8mb4_unicode_ci COMMENT '
优惠政策',
      'is_hot' varchar(2) CHARACTER SET utf8mb4 COLLATE utf8mb4_unicode_ci DEFAULT '2' 
COMMENT '是否热门楼盘 1 是,2 否',
      PRIMARY KEY ('id') USING BTREE,
      KEY 'FKk5dgddy7507itba3soim8otj8' ('developer_info_id') USING BTREE,
      KEY 'FKslc4mjlrebudyhljut9b2bjad' ('community_id') USING BTREE,
      KEY 'FKf4550hdwqte1ljcc3ny0673ws' ('property_info_id') USING BTREE,
      CONSTRAINT 'FKf4550hdwqte1ljcc3ny0673ws' FOREIGN KEY ('property_info_id') 
REFERENCES 'service_property_info' ('id') ON DELETE RESTRICT ON UPDATE RESTRICT,
      CONSTRAINT 'FKk5dgddy7507itba3soim8otj8' FOREIGN KEY ('developer_info_id') 
REFERENCES 'service_developer_info' ('id') ON DELETE RESTRICT ON UPDATE RESTRICT,
      CONSTRAINT 'FKslc4mjlrebudyhljut9b2bjad' FOREIGN KEY ('community_id') 
REFERENCES 'service_community_info' ('id') ON DELETE RESTRICT ON UPDATE RESTRICT
    ) ENGINE=InnoDB DEFAULT CHARSET=utf8mb4 COLLATE=utf8mb4_unicode_ci 
ROW_FORMAT=DYNAMIC;
```

（5）销售信息表，表名service_sale_info，用于管理销售信息。

```sql
    CREATE TABLE 'service_sale_info' (
      'id' varchar(32) CHARACTER SET utf8mb4 COLLATE utf8mb4_unicode_ci NOT NULL,
      'create_date' datetime(6) DEFAULT NULL,
      'del_flag' int NOT NULL,
      'update_date' datetime(6) DEFAULT NULL,
      'build' varchar(255) CHARACTER SET utf8mb4 COLLATE utf8mb4_unicode_ci DEFAULT 
'' COMMENT '楼栋',
      'house' varchar(255) CHARACTER SET utf8mb4 COLLATE utf8mb4_unicode_ci DEFAULT 
'' COMMENT '户名',
```

```
        'id_card' varchar(255) CHARACTER SET utf8mb4 COLLATE utf8mb4_unicode_ci DEFAULT
'' COMMENT '身份证号',
        'img_url' varchar(2000) CHARACTER SET utf8mb4 COLLATE utf8mb4_unicode_ci DEFAULT
'' COMMENT '合同附件',
        'name' varchar(255) CHARACTER SET utf8mb4 COLLATE utf8mb4_unicode_ci DEFAULT ''
COMMENT '购买者姓名',
        'phone' varchar(255) CHARACTER SET utf8mb4 COLLATE utf8mb4_unicode_ci DEFAULT
'' COMMENT '手机号',
        'price' varchar(255) CHARACTER SET utf8mb4 COLLATE utf8mb4_unicode_ci DEFAULT
'' COMMENT '成交价',
        'unit' varchar(255) CHARACTER SET utf8mb4 COLLATE utf8mb4_unicode_ci DEFAULT ''
COMMENT '单元',
        'community_id' varchar(32) CHARACTER SET utf8mb4 COLLATE utf8mb4_unicode_ci
DEFAULT NULL,
        'user_id' varchar(32) CHARACTER SET utf8mb4 COLLATE utf8mb4_unicode_ci DEFAULT
NULL,
        PRIMARY KEY ('id') USING BTREE,
        KEY 'FK9lm3m538yti3drusnrk2s05d2' ('community_id') USING BTREE,
        KEY 'FKjc2nbaxqypuj01ga2q43ntfq' ('user_id') USING BTREE,
        CONSTRAINT 'FK9lm3m538yti3drusnrk2s05d2' FOREIGN KEY ('community_id')
REFERENCES 'service_community_info' ('id') ON DELETE RESTRICT ON UPDATE RESTRICT,
        CONSTRAINT 'FKjc2nbaxqypuj01ga2q43ntfq' FOREIGN KEY ('user_id') REFERENCES
'sys_user' ('id') ON DELETE RESTRICT ON UPDATE RESTRICT
    ) ENGINE=InnoDB DEFAULT CHARSET=utf8mb4 COLLATE=utf8mb4_unicode_ci
ROW_FORMAT=DYNAMIC;
```

3.2.4 系统文件说明

本项目系统文件的组织结构如图3.12和图3.13所示。

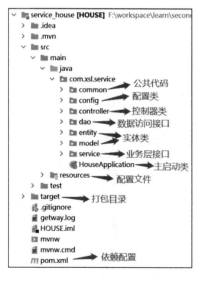

图3.12 项目后端文件组织结构

图3.13 项目前端文件组织结构

下面对项目后端文件各个部分作简要说明：

（1）common：包含一些常用的公用类，例如日期处理、HTTP工具、常量维护等。

（2）config：Java配置类，配置一些不在配置文件中的信息，例如拦截器配置、文件上传参数配置等。

（3）controller：控制器类，所有controller层的类。

（4）实体类：entity是与数据库对应的实体类；model里面包括em和vo。em是返回封装对象；vo是数据映射对象，用于页面展示。

（5）dao：使用MyBatis实现的数据访问层接口。

（6）service：业务逻辑的主要实现层，在该层可以实现事务。

（7）resources：包括Spring Boot的配置文件，这里使用的是application.yml形式，还包括一些基础参数配置config.properties。

（8）pom.xml：依赖JAR包，使用Maven进行项目构建时，依赖JAR包统一在pom文件中进行管理。

（9）target：打包目录，构建完成后存放项目war包或JAR包的目录。

下面对项目前端文件各个部分作简要说明：

（1）assets：包含图片或文件资源。
（2）components：所有页面。
（3）router：页面路由，指定页面跳转。
（4）utils：工具，主要有HTTP工具。
（5）vue.config.js：配置文件。

3.3 项目前端设计

3.3.1 登录

用户输入正确的用户名和密码，即可完成登录。登录页面如图3.14所示。

图3.14 登录页面

以下是登录页面的代码，使用Vue的模板语法构造页面。

```html
<template>
    <div class="login-wrap">
        <div class="ms-login loginForm">
            <div class="ms-title"><a href="#" style="color: #46ABF1">二手房系统</a></div>
            <el-form :model="param" :rules="rules" ref="login" label-width="0px" class="ms-content">
                <el-form-item prop="username">
                    <section class="loginForm-input">
                        <el-input v-model="param.userName" placeholder="请输入账号" calss="loginForm-el-input">
                        </el-input>
                    </section>
                </el-form-item>
                <el-form-item prop="password">
                    <el-input calss="loginForm-el-input"
                        type="password"
                        placeholder="请输入密码"
                        v-model="param.password"
                        @keyup.enter.native="submitForm()">
                    </el-input>
                </el-form-item>
                <div class="login-btn">
                    <el-button type="primary " @click="submitForm()" round>登录</el-button>
                </div>
                <p class="login-tips">Tips：初始密码为手机号后6位。</p>
            </el-form>
        </div>
    </div>
</template>
```

语句块<template>是一种模板占位符，每个*.vue 文件最多可同时包含一个顶层 <template>块，其中的内容会被提取出来并传递给 @vue/compiler-dom，预编译为 JavaScript 的渲染函数，并附属到导出的组件上作为其 render 选项。在<template>语句块中，可以使用Vue提供的各种模板语句。

<el-form>是Element UI提供的form组件。除了form组件外，Element UI还提供了其他常用的各种组件，详细信息可以参考官方API文档。

在上述页面中用到了v-model、:model、:rules、@keyup.enter.native、@click等Vue指令。v-前缀作为一种视觉提示，用来识别模板中Vue特定的属性。当我们在使用Vue.js为现有标签添加动态行为时，v- 前缀很有帮助，然而，对于一些频繁用到的指令来说，这种方式很烦琐。所以常用的一些指令可以使用缩写命令，其中，model是v-model的缩写，:rules是v-rules的缩写，@click是v-on:click的缩写，@keyup.enter.native是v-on:keyup.enter.native的缩写。

v-model用来在表单 <input>、<textarea> 及 <select> 元素上创建双向数据绑定。例如，页面中的账号输入框,它会根据控件类型自动选取正确的方法来更新元素，负责监听用户的输入事件以更新数据,并对一些极端场景进行一些特殊处理。v-model会忽略所有表单元素的value、checked、selected attribute的初始值,而将Vue实例的数据作为数据来源。可以通过JavaScript在组件的data选项中声明初始值,这个初始值可以是静态的默认值,也可以是通过异步请求返回而刷新的数据值。

v-model还支持修饰符,常用的有3种,如表3.5所示。

表3.5　v-model常用修饰符

修饰符名称	用　　法	作　　用
.lazy	`<input v-model.lazy="msg">`	在change时而非input时更新
.number	`<input v-model.number="age" type="number">`	自动将用户的输入值转为数值类型
.trim	`<input v-model.trim="msg">`	自动过滤用户输入的首尾空白字符

除了v-model外,Vue还有其他指令同样使用v-前缀来标识。常用的有用于数据绑定的v-bind,用于事件绑定的v-on,用于条件渲染的v-if,每个指令的使用方法和作用可以查看Vue的官方文档。

rules指令可以用来进行输入框验证。具体的动作和返回可以查看对应的JS方法,示例代码如下:

```
rules: {
    userName: [{ required: true, message: '请输入账号', trigger: 'blur' }],
    password: [{ required: true, message: '请输入密码', trigger: 'blur' }]
}
```

登录按钮中,使用@click="submitForm()"绑定单击事件,在JS中定义该方法,处理登录以及登录后的回调方法:

```
submitForm() {
    this.$http({
        method: 'post',
        url: '/user/login',                              //登录接口地址
        data: this.param
    }).then((res) => {
        if (res.data.status === 200) {
            var roleType = res.data.data.roleType;       //角色类型
            var username = res.data.data.name;           //用户名
            sessionStorage.setItem('userId', res.data.data.id);   //将用户id放入session
            sessionStorage.setItem('ms_username', username);
            sessionStorage.setItem('roleType', roleType);
            if (roleType != null) {
                if (this.managePower.indexOf(roleType) > -1) {
                    this.$message.success('登录成功');     //有管理权限
                    this.$router.push('/');
                } else {
                    this.$message.error('没有权限');
                }
            } else {
                this.$message.error('登录失败');
            }
        } else {
```

```
            this.$message.error(res.data.message);
        }
    }, (err) => {
        console.log(err);
    });
}
```

3.3.2 二手房房源管理

二手房房源管理主要涉及二手房房源信息的搜索、新增、修改和删除。二手房房源管理页面如图3.15所示。

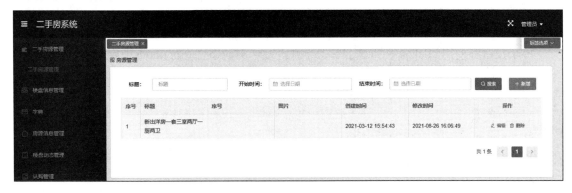

图3.15 二手房房源管理页面

页面中时间组件的代码如下：

```
<el-table-column prop="createDate" label="创建时间" :formatter="timeStamp2String">
</el-table-column>
<el-table-column prop="updateDate" label="修改时间" :formatter="timeStamp2String">
</el-table-column>
```

这里的查询条件中用到了:formatter，它是v-formatter的缩写。这个命令可以用来格式化时间。timeStamp2String是自定义的日期转换函数，这个函数定义在methods中。

```
//在Jquery里格式化Date日期时间数据
timeStamp2String(row, column, cellValue) {
    let time = cellValue;
    var dateee = new Date(time).toJSON();
    var date = new Date(+new Date(dateee) + 8 * 3600 * 1000).toISOString().replace(/T/g, ' ').replace(/\.[\d]{3}Z/, '');
    return date;
},
//在Jquery里格式化Date日期时间数据
dateStamp2String(row, column, cellValue) {
    let time = cellValue;
    var dateee = new Date(time).toJSON();
    var date = new Date(+new Date(dateee) + 8 * 3600 * 1000).toISOString().replace(/T/g, ' ').replace(/\.[\d]{3}Z/, '');
    return date.substr(0, 10);
},
```

除了以上函数外，methods内还一一定义了编辑、添加、搜索、删除按钮对应的方法，代码如下：

```
    <el-button type="primary" icon="el-icon-search" class="handle-del mr10 "
@click="handleSearch">搜索
    </el-button>
    <el-button type="success" icon="el-icon-plus" class="handle-del mr10"
@click="handleAdd">新增</el-button>
    <el-button type="text" icon="el-icon-edit" @click="handleEdit(scope.$index,
scope.row)">编辑
    </el-button>
    <el-button type="text" icon="el-icon-delete" class="red"
@click="handleDelete(scope.$index, scope.row)" >删除
    </el-button>
```

每个按钮都使用:click命令绑定了单击事件，逐一对应事件在methods中定义的方法。

```
    // 触发搜索按钮
    handleSearch() {
        ...
    },
    // 删除操作
    handleDelete(index, row) {
    ...
    },
    // 多选操作
    handleSelectionChange(val) {
        ...
    },
    handleAdd() {
        ...
    },
    // 编辑操作
    handleEdit(index, row) {
        ...
    }
```

handleSearch是单击"搜索"按钮时调用的方法。handleAdd是单击"新增"按钮时调用的方法。handleEdit是单击"编辑"按钮时调用的方法，会通过index和row参数把当前对应行数据传递过来。handleDelete是单击"删除"按钮时调用的方法，删除成功后要重新刷新该列表页面。

3.3.3 楼盘信息管理

楼盘信息管理与前面的二手房房源管理类似，包括楼盘信息的搜索、新增、修改和删除。楼盘信息管理页面如图3.16所示。

由于楼盘信息比较复杂，在列表中不能直观地全部显示出来，所以在列表最左侧加上展开按钮，单击展开按钮后可以查看详细信息。楼盘详细信息页面如图3.17所示。

展开按钮对应的部分代码如下：

```
<el-table-column type="expand">
    <template>
        ...//此处为每一行缩放内容
    </template>
```

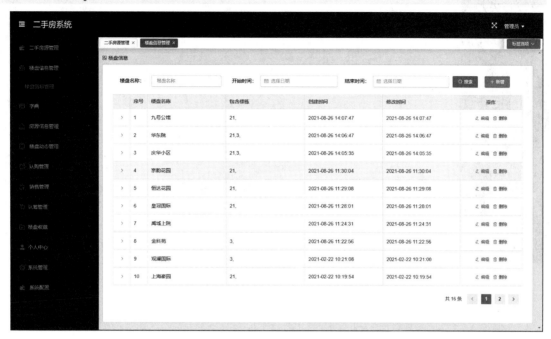

图3.16 楼盘信息管理页面

图3.17 楼盘详细信息页面

使用<template>实现左侧展开按钮内容。

添加楼盘信息时，单击上方的"新增"按钮，打开新增楼盘页面，如图3.18所示。

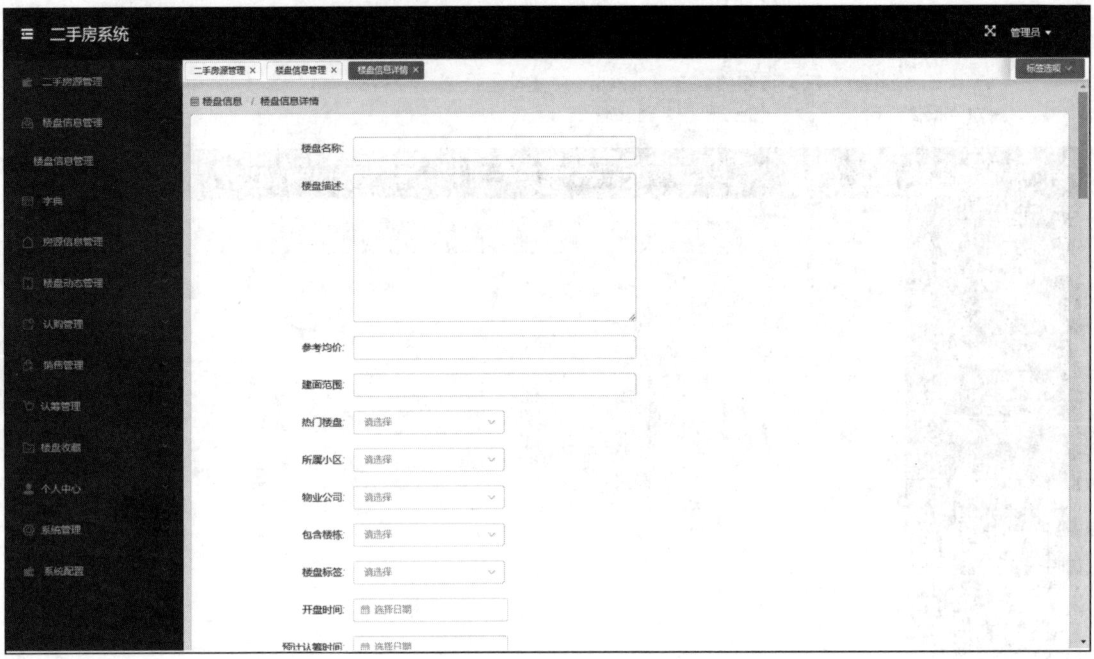

图3.18　新增楼盘信息页面

对应的代码如下：

```
<!-- 新增弹出框 -->
<el-dialog title="新增" :visible.sync="addVisible" width="30%" ref="formAdd">
    ...
</el-dialog>
```

这里使用:visible.sync动态控制页面的显示与隐藏。

除了普通信息外，还需要上传相关图片，例如VR图片、现场图片、样板间图片等，对应代码如下：

```
<el-upload :action="this.global.fileUpload" list-type="picture-card"
    :file-list="imgList" :on-success="handleFile" :on-preview="handlePreview"
    :on-remove="handleRemoveFile">
    <i class="el-icon-plus"></i>
</el-upload>
```

这里用到了上传组件el-upload。action属性配置的是上传的接口地址；on-success设置的是文件上传成功后的回调方法，在这里可以将上传成功后的图片显示在页面中；on-preview设置的是单击文件列表中已上传文件时的方法；on-remove设置的是移除文件时执行的方法。对应的JS方法如下：

```
//图片的3个方法
handleFile(file, fileList) {
    if (this.form.imgUrl == undefined) {
        this.form.imgUrl = fileList.response.data.url + ',';
    } else {
        this.form.imgUrl += fileList.response.data.url + ',';
```

```
            }
            if (this.form.imgUrl.length > 0) {
                this.form.imgUrl = this.form.imgUrl.replace('undefined', '');
                this.form.imgUrl = this.form.imgUrl.replace('null', '');
            }
        },
        handlePreview(file) {
            this.dialogImgUrl = file.url;
            this.dialogVisible = true;
        },
        handleRemoveFile(file, fileList) {
            if (file.response == undefined) {
                let path = this.global.uploadPath ;
                let name = file.url.replace(path, '');

                this.form.imgUrl = this.form.imgUrl.replace(name + ',', '');
            } else {
                let url = file.response.data.url;
                this.form.imgUrl = this.form.imgUrl.replace(url + ',', '');
            }
        }
```

上传成功后，将返回的文件链接地址设置到请求参数中，通过handlePreview将图片显示出来；移除文件时，删除请求参数中该图片对应的信息。

3.3.4 房源信息管理

房源信息管理是指所有房源信息的管理，与二手房不同的是，这些房源是由开发商直接提供给购买者，也可以理解为新楼盘房源信息。新楼盘售出后，业主再计划卖出的房源会被归为二手房房源。房源信息管理页面如图3.19所示。

图3.19 房源信息管理页面

单击"新增"或者"编辑"按钮就会触发对应的方法，弹出如图3.20所示的页面。

这里的小区列表和户型选择等下拉框列表数据通过异步调用接口来获取，例如JS代码中的getCommunityList方法和getHouseTypeInfo方法。

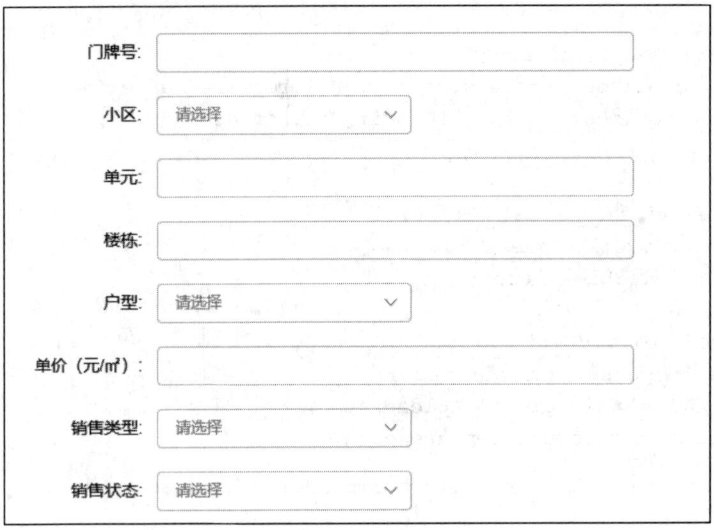

图3.20 新增或修改房源信息

3.3.5 楼盘动态管理

楼盘动态管理主要用来维护楼盘动态信息。使用者可在后台便捷地操作需要公布的楼盘信息，例如销售状态、楼盘建设进度等。这些操作包括楼盘动态的查询、新增、修改和删除。楼盘动态管理页面如图3.21所示。

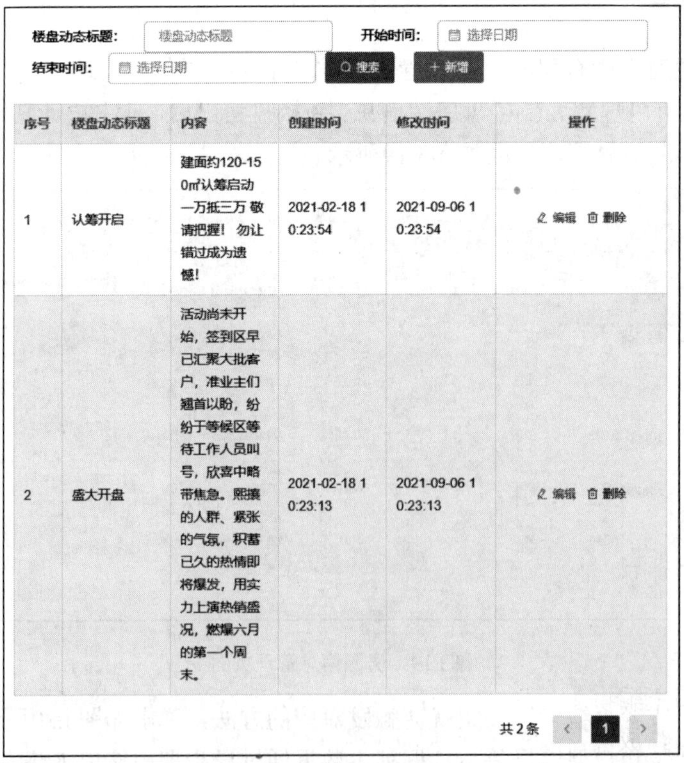

图3.21 楼盘动态管理页面

新增楼盘动态页面如图3.22所示。

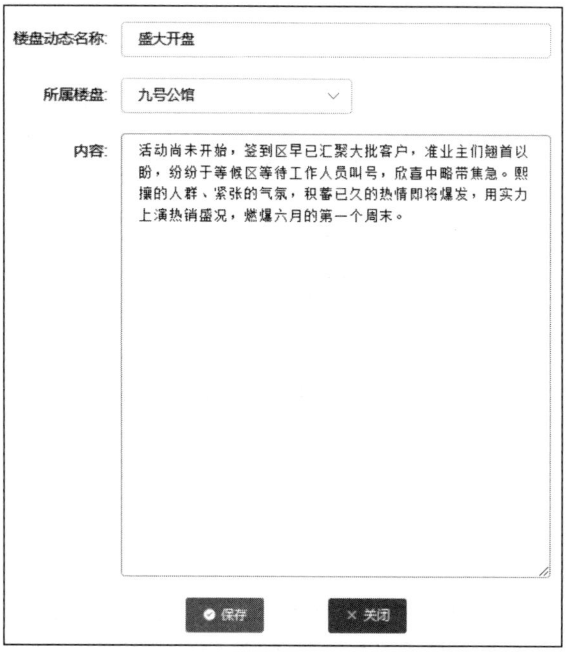

图3.22　新增楼盘动态页面

3.3.6　认购管理

认购管理主要对客户认购进行管理。认购记录的生成在App客户端或其他调用生成认购记录的终端上进行。因为房产购买需要通过客户操作和销售人员确认，所以后台只包括认购记录支付状态的维护以及认购记录的删除。认购管理页面如图3.23所示。

图3.23　认购管理页面

在该页面单击"编辑"按钮，弹出如图3.24所示的编辑页面。修改完成后单击"保存"按钮，调用保存接口，成功保存后执行回调方法，刷新列表页。

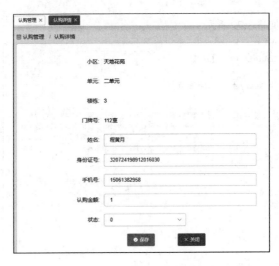

图3.24 修改认购记录

3.3.7 销售管理

销售管理主要对订单进行管理,包括搜索、新增、修改和删除销售记录。除了管理订单的基本信息外,销售合同将作为图片附件一起上传。销售管理页面如图3.25所示。

在该页面上单击"新增"或者"编辑"按钮,会弹出新增或编辑对话框如图3.26所示。

图3.25 销售管理页面

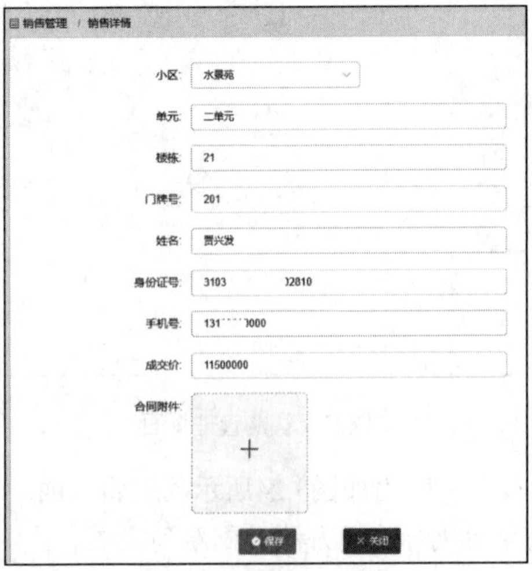

图3.26 新增或修改销售记录

3.3.8 认筹管理

认筹是指房地产开发商在取得预售证后但未被批准正式销售之前进行的活动。购房者预交一定的认筹金,可以获得购房优惠或优先选择房源等。认筹管理主要对认筹记录和认筹者进行管理,主要操作包括搜索、新增、修改和删除等。列表页面如图3.27、图3.28所示。

图3.27 认筹管理页面

图3.28 认筹者管理页面

认筹者管理功能相对简单,是为了更直观、更方便地对认筹进行管理而设置的功能。由于篇幅有限,不再详述。主要看认筹管理部分。

在该页面上单击"新增"或者"编辑"按钮,会弹出新增或编辑的对话框,如图3.29所示。

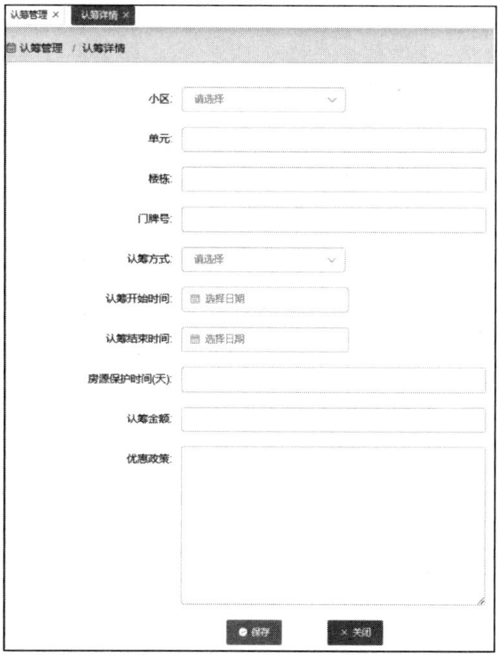

图3.29 新增或修改认筹记录

3.3.9 楼盘收藏管理

用户在浏览App或Web终端时，会将一些意向楼盘或房源放入收藏夹，便于及时关注房源消息。楼盘收藏管理就是查看和搜索收藏记录，以便为用户精准推送楼盘或房源消息，使用户获取到更有用的讯息。楼盘收藏管理页面如图3.30所示。

图3.30 楼盘收藏管理页面

3.3.10 系统管理与系统设置

从用户管理往后基本都是系统设置和系统管理的内容，所使用的组件和前端技术在之前的模块中讲解过，所以后面的功能页面只简单介绍下功能，不再详细介绍具体实现，读者可根据本书配套源码自行研究。

1. 用户管理

用户管理用于管理后台系统登录用户以及为用户设置角色。用户管理页面如图3.31所示。

图3.31 用户管理页面

2. 角色管理

角色管理用来区分用户的角色，可用于实现权限控制，主要包括角色的新增、修改和删除。角色管理页面如图3.32所示。

图3.32 角色管理页面

3. 顾问管理

顾问管理用来实现销售顾问的管理。这里包括在App或其他终端通过接口注册或添加的顾问，也包括后台添加的顾问。除了能查看基本信息外，还能查看顾问的头像和微信二维码。顾问管理页面如图3.33所示。

图3.33 顾问管理页面

4. 户型管理

户型管理用来实现房源和楼盘所对应的户型的管理。在这里对户型信息进行维护后，其他模块在使用户型信息时就可以动态刷新户型数据。户型管理页面如图3.34所示。

5. 装修类型管理

装修类型包括精装修、简装修、普通装修、毛坯等。装修类型管理页面如图3.35所示。

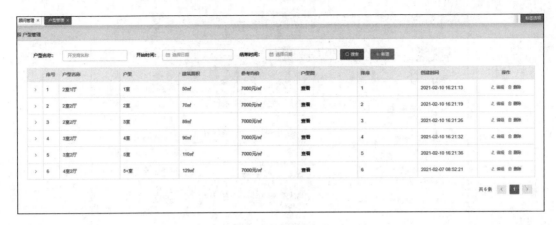

图3.34 户型管理页面

图3.35 装修类型管理页面

6. 建筑类型管理

建筑类型包括高层、小高层、洋房、多层、联排别墅、独栋别墅、复式等。建筑类型管理页面如图3.36所示。

图3.36 建筑类型管理页面

7. 建筑标签管理

建筑标签包括海景房、商业街、医院旁、公园旁、高绿化、中式园林、欧式园林等。建筑标签管理页面如图3.37所示。

8. 物业类型管理

物业类型包括住宅、写字楼、商铺、别墅等。物业类型管理页面如图3.38所示。

图3.37 建筑标签管理页面

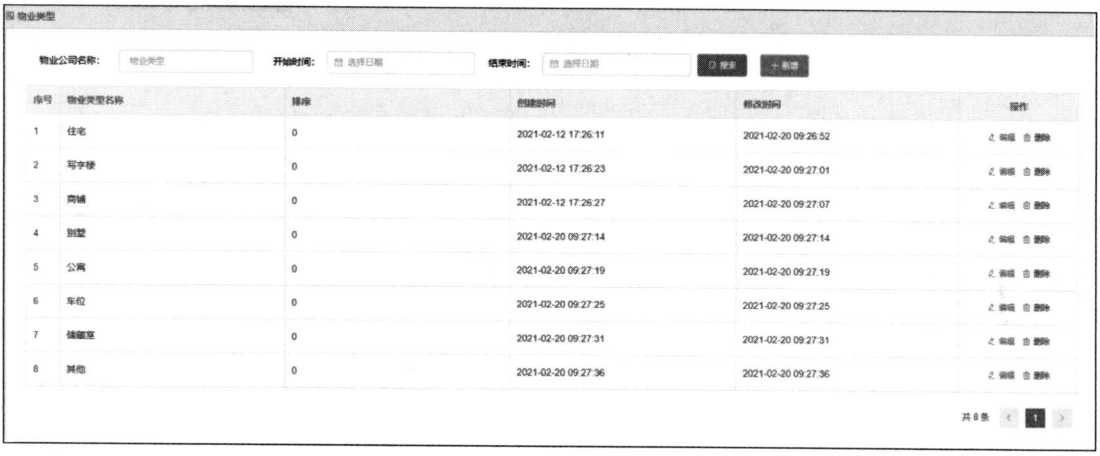

图3.38 物业类型管理页面

9. 优惠券管理

优惠券是客户能够使用的优惠。优惠券管理页面如图3.39所示。

图3.39 优惠券管理页面

10. 开发商管理

开发商管理是对所有开发商公司进行管理,用于房源或楼盘信息展示。开发商管理页面如图3.40所示。

图3.40　开发商管理页面

11. 物业公司管理

物业公司管理是对物业公司进行管理，用于房源或楼盘信息展示。物业公司管理页面如图3.41所示。

图3.41　物业公司管理页面

12. 小区管理

小区管理是对小区进行管理，用于房源或楼盘信息展示、销售顾问关联信息等。小区管理页面如图3.42所示。

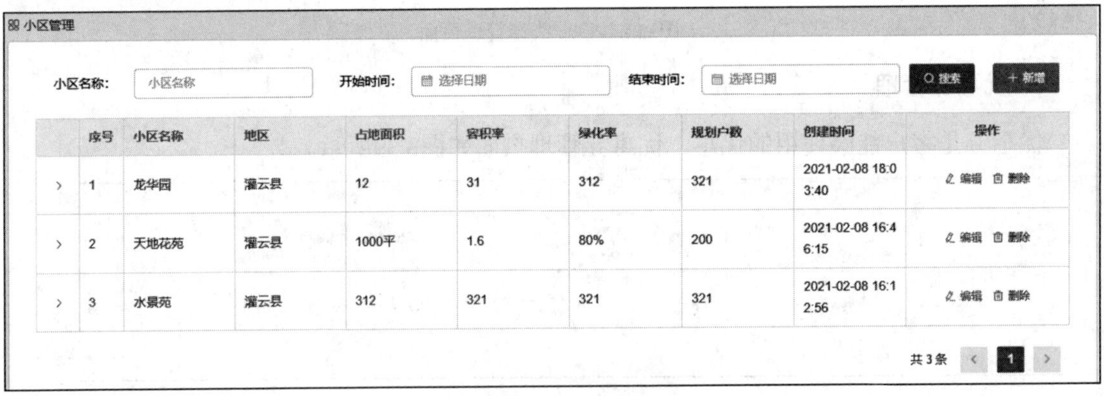

图3.42　小区管理页面

3.4 项目后端实现

3.4.1 通用类

1. 分页类PageModel

本项目中所有需要分页的数据,其返回结果统一使用PageModel类进行封装。与上一章不同的是,本章的PageModel采用了泛型结构,将分页返回的数据一并封装到了该类中。

```java
//分页类
public class PageModel<E> implements Serializable {
    private static final long serialVersionUID = 3265524976080127173L;
    private int totalCount;                 //总记录数
    private int pageSize = 10;              //每页显示的数量
    private int totalPage;                  //总页数
    private int currentPage = 1;            //当前页数
    private List<E> list;                   //分页集合列表
    private String url;                     //分页跳转的URL
    public PageModel() {
    }
    public PageModel(Page page) {
        this.totalCount = Integer.parseInt(page.getTotalElements() + "");
        this.pageSize = page.getSize();
        this.totalPage = page.getTotalPages();
        this.list = page.getContent();
        //page.get
        this.currentPage = page.getNumber();
    }
    public PageModel(int pageSize) {
        this.pageSize = pageSize;
    }
    public int getTotalCount() {
        return totalCount;
    }
    public void setTotalCount(int totalCount) {
        this.totalCount = totalCount;
    }
    public int getPageSize() {
        return pageSize;
    }
    public void setPageSize(int pageSize) {
        this.pageSize = pageSize;
    }
    public int getTotalPage() {
        return totalPage;
    }
    public void setTotalPage(int totalPage) {
```

```java
        this.totalPage = totalPage;
    }
    public int getCurrentPage() {
        return currentPage;
    }
    public void setCurrentPage(int currentPage) {
        this.currentPage = currentPage;
    }
    public List<E> getList() {
        return list;
    }
    public void setList(List<E> list) {
        this.list = list;
    }
    public String getUrl() {
        return url;
    }
    public void setUrl(String url) {
        this.url = url;
    }
}
```

2. 返回结果封装类Result

封装结果包括响应码、提示消息、响应数据等。

```java
@ApiModel("统一响应模型")
public class Result {
    @ApiModelProperty("响应状态")
    private int status;
    @ApiModelProperty("响应消息")
    private String message;
    @ApiModelProperty("响应数据")
    private Object data;
    @ApiModelProperty("总数")
    private Object count;
    public Object getCount() {
        return count;
    }
    public void setCount(Object count) {
        this.count = count;
    }
    public Result setStatus(ResultStatusEnum resultStatus) {
        this.status = resultStatus.getStatus();
        return this;
    }
    public int getStatus() {
        return status;
    }
    public Result setStatus(int status) {
        this.status = status;
        return this;
```

```java
    }
    public String getMessage() {
        return message;
    }
    public Result setMessage(String message) {
        this.message = message;
        return this;
    }
    public Object getData() {
        return data;
    }
    public Result setData(Object data) {
        this.data = data;
        return this;
    }
}
```

3. 通用返回码ResultStatusEnum

封装返回结果中的状态码。

```java
public enum ResultStatusEnum {
    //成功
    SUCCESS(200),
    //失败
    FAIL(201),
    //未认证（签名错误）
    UNAUTHORIZED(401),
    //接口不存在
    NOT_FOUND(404),
    //服务器内部错误
    INTERNAL_SERVER_ERROR(500);
    private int status;
    ResultStatusEnum(int status) {
        this.status = status;
    }
    public int getStatus() {
        return status;
    }
}
```

4. 通用查询请求参数CommonQueryDTO

封装常用的查询所需的参数，包括当前页、每页数量、查询日期等。

```java
@Data
public class CommonQueryDTO {
    private Integer currentPage;
    private Integer pageSize;
    private String delFlag;
    private String id;
    private String startDate;
    private String endDate;
```

```java
    private int sort;
}
```

5. 通用数据库操作类CommonDao以及实现类CustomBaseSqlDaoImpl

```java
@NoRepositoryBean
public interface CommonDao<E, ID extends Serializable> extends JpaRepository<E, ID>, JpaSpecificationExecutor<E> {
}
public class CustomBaseSqlDaoImpl {
    @Autowired
    private EntityManager em;
    public List<Map<String, Object>> querySqlObjects(String sql, Integer currentPage, Integer rowsInPage) {
        return this.querySqlObjects(sql, null, currentPage, rowsInPage);
    }
    public List<Map<String, Object>> querySqlObjects(String sql) {
        return this.querySqlObjects(sql, null, null, null);
    }
    public List<Map<String, Object>> querySqlObjects(String sql, List<Object> params) {
        return this.querySqlObjects(sql, params, null, null);
    }
    @SuppressWarnings("unchecked")
    public List<Map<String, Object>> querySqlObjects(String sql, Object params, Integer currentPage, Integer rowsInPage) {
        Query qry = em.createNativeQuery(sql);
        SQLQuery s = qry.unwrap(SQLQuery.class);
        //设置参数
        if (params != null) {
            if (params instanceof List) {
                List<Object> paramList = (List<Object>) params;
                for (int i = 0, size = paramList.size(); i < size; i++) {
                    qry.setParameter(i + 1, paramList.get(i));
                }
            } else if (params instanceof Map) {
                Map<String, Object> paramMap = (Map<String, Object>) params;
                for (String key : paramMap.keySet()) {
                    qry.setParameter(key, paramMap.get(key));
                }
            }
        }

        if (currentPage != null && rowsInPage != null) {          //判断是否有分页
            // 起始对象位置
            qry.setFirstResult(rowsInPage * (currentPage - 1));
            // 查询对象个数
            qry.setMaxResults(rowsInPage);
        }
        s.setResultTransformer(CriteriaSpecification.ALIAS_TO_ENTITY_MAP);
        List<Map<String, Object>> resultList = new ArrayList<Map<String, Object>>();
```

```java
        try {
            resultList = s.list();
        } catch (Exception e) {
        } finally {
            em.close();
        }
        return resultList;
    }
    ...
}
```

3.4.2 登录

根据请求传递过来的用户名和密码去数据库比对用户表sys_user，如果用户名和密码正确，则将用户放入session中，返回登录成功；否则，返回登录失败。

```java
@Override
public Result login(HttpSession session, HttpServletRequest request, User user) {
    String password = "";
    password = Base64.getEncoder().encodeToString(user.getPassword().getBytes(StandardCharsets.UTF_8));
    List<User> list = userRepository.getUserList(user.getUserName(), password);
    //根据用户名和密码查询用户
    User userInfo = new User();
    if (list != null && list.size() > 0) {
        userInfo = list.get(0);
        UserContext.putCurrentUser(userInfo);                   //放入session
    }
    if (userInfo == null) {
        return ResultGenerator.genFailResult("登录失败");
    } else {
        return ResultGenerator.genSuccessResult(userInfo);
    }
}
```

3.4.3 二手房房源管理

（1）查询房源

查询房源时，可以通过分页查询，也可以单独查询某个房源的具体信息。分页查询时，查询条件包括房源名称、创建时间。

```java
@Override
public PageModel<TwoHandHouseInfo> queryInfoPage(TwoHandHouseInfoModel dto) {
    Map<String, Object> map = new HashMap<String, Object>();
    StringBuilder hql = new StringBuilder();                    //SQL构建
    hql.append(" select t from TwoHandHouseInfo t where t.delFlag ='0' ");
    if (dto != null) {
        String title = dto.getTitle();
        if (StringUtils.isNotBlank(title)) {
            hql.append(" and t.title like :title ");            //SQL拼接
```

```
            map.put("title", "%" + title + "%");              //参数传递
        }
        String createTime = dto.getStartDate();
        if (StringUtils.isNotBlank(createTime)) {
            hql.append(" and t.createDate >= :createTime ");   //SQL拼接
            map.put("createTime", createTime);                 //参数传递
        }
        String endTime = dto.getEndDate();
        if (StringUtils.isNotBlank(endTime)) {
            hql.append(" and t.createDate <= :endTime ");      //SQL拼接
            map.put("endTime", endTime);//参数传递
        }
    }
    hql.append(" order by t.createDate desc ");
    return this.queryForPageWithParams(hql.toString(), map, dto.getCurrentPage(),
dto.getPageSize());
}
```

构造完查询条件之后,通过CustomBaseSqlDaoImpl中的公用查询方法queryForPageWithParams实现数据库查询。

(2) 新增或修改房源信息

根据参数中是否存在id来判断,如果不存在则为新增,如果存在则为修改。

```
@Override
public Result save(TwoHandHouseInfoModel model) {
    TwoHandHouseInfo preInfo = new TwoHandHouseInfo();
    String id = model.getId();
    if (!StringUtils.isNullOrEmpty(id)) {
        TwoHandHouseInfo info = repository.findFirstById(id);
        preInfo.setCreateDate(info.getCreateDate());
    }
    BeanUtils.copyProperties(model, preInfo);
    return ResultGenerator.genSuccessResult(repository.save(preInfo));
}
```

(3) 删除房源信息

删除为逻辑删除,即修delFlag字段的值为1。

```
@Override
public Result delete(TwoHandHouseInfoModel model) {
    try {
        repository.delete(model.getDelFlag(), model.getId());
        return ResultGenerator.genSuccessResult("删除成功");
    } catch (Exception e) {
        e.printStackTrace();
        return ResultGenerator.genFailResult("删除失败");
    }
}
```

repository中的实现方法如下:

```
@Modifying
@Transactional
@Query(value = "update two_hand_house_info set del_flag =?1 where id =?2",
nativeQuery = true)
void delete(String delFlag, String id);
```

在方法上方使用@Query注解设置操作语句，在操作语句中通过占位符"?"传递参数，解析SQL语句时将占位符替换为实际的参数值，完成SQL语句的拼接。"?"后面的序号代表参数的顺序，与方法中的参数一一对应。在上述delete方法中，对应的SQL语句中"?1"所对应的就是delFlag的参数值，"?2"对应的就是id的参数值。

3.4.4 楼盘信息管理

（1）查询楼盘信息

构造完查询条件之后，直接调用repository中的方法获取查询结果。

```
@Query(value = "select * from service_house_info where del_flag='0' " +
        " and (title like ?1 or '%%' =?1)" +
        " and (create_date >=?2 or ''=?2)" +
        " and (create_date <=?3 or ''=?3)" +
        " order by create_date desc limit ?4,?5 ", nativeQuery = true)
List<HouseInfo> pageList(String title, String startDate, String endDate, int
currentPage, int pageSize);

@Query(value = "select count(1) from service_house_info where del_flag='0'" +
        " and (title like ?1 or '%%' =?1)" +
        " and (create_date >=?2 or ''=?2)" +
        " and (create_date <=?3 or ''=?3)", nativeQuery = true)
int countPageList(String title, String startDate, String endDate);
```

使用@Query注解设置操作语句，通过占位符传递参数。

（2）添加或修改楼盘信息

```
@Override
public Result save(RealEstateModel model) {
    RealEstate preInfo = new RealEstate();
    String id = model.getId();
    if (!StringUtils.isEmpty(id)) {                          //判断id是否为空
        RealEstate info = repository.findFirstById(id);      //不为空则修改
        preInfo.setCreateDate(info.getCreateDate());
    }
    BeanUtils.copyProperties(model, preInfo);
    return ResultGenerator.genSuccessResult(repository.save(preInfo));
}
```

（3）查询单个楼盘信息

```
@Override
public Result info(RealEstateModel model) {
    RealEstate realEstate = repository.findFirstById(model.getId());//楼盘基本信息
```

```java
        String realEstateDecorationInfoTitle = "";
        List<DecorationInfo> decorationList = realEstate.getDecorationInfos();
        realEstate.setHouseType(getTitle(realEstate));
        if (decorationList != null && decorationList.size() > 0) {   //装修信息查询
            String decorationTitle = decorationList.get(0).getTitle();
            if (decorationTitle.equals("毛坯")) {
                realEstateDecorationInfoTitle = decorationTitle;
            } else {
                realEstateDecorationInfoTitle = "带装修";
            }
        }
        String communityId = realEstate.getCommunity().getId();      //小区id
        List<Consultant> consultantList =
consultantRepository.findAllByCommunityId(communityId);
        realEstate.setConsultants(consultantList);                   //顾问列表
        String buildingInfos = realEstate.getBuildingInfos();
        if (buildingInfos.endsWith(",")) {                           //建筑信息
            buildingInfos = buildingInfos.substring(0, buildingInfos.length() - 1);
            realEstate.setRealEstateDecorationInfoTitle(realEstateDecorationInfoTitle);
        }
        String userId = StringUtils.isNotBlank(model.getUserId()) ? model.getUserId() : "";
        realEstate.setUserRealEstateFlag("2");
        realEstate.setUserRealEstateFlagBj("2");
        realEstate.setUserRealEstateFlagSc("2");
        if (!"".equals(userId)) {
            //开盘提醒
            List<UserRealEstate> userRealEstates = userRealEstateRepository.
userRealEstateList(realEstate.getId(), userId,"2");
            if (userRealEstates != null && userRealEstates.size() > 0) {
                realEstate.setUserRealEstateFlag("1");
                realEstate.setUserRealEstateId(userRealEstates.get(0).getId());
            }
            // 变价通知
            List<UserRealEstate> userRealEstatesBj = userRealEstateRepository.
userRealEstateList(realEstate.getId(), userId,"1");
            if (userRealEstatesBj != null && userRealEstatesBj.size() > 0) {
                realEstate.setUserRealEstateFlagBj("1");
                realEstate.setUserRealEstateBjId(userRealEstatesBj.get(0).getId());
            }
            //收藏
            List<UserRealEstate> userRealEstatesSc = userRealEstateRepository.
userRealEstateList(realEstate.getId(), userId,"3");
            if (userRealEstatesSc != null && userRealEstatesSc.size() > 0) {
                realEstate.setUserRealEstateFlagSc("1");
                realEstate.setUserRealEstateScId(userRealEstatesSc.get(0).getId());
            }
        }
        return ResultGenerator.genSuccessResult(realEstate);
    }
```

根据id查询出某个楼盘信息后，要对其中的数据作转换，例如装修情况、置业顾问、建筑信息等。

3.4.5 房源信息管理

（1）查询房源

```
@Override
public PageModel<HouseInfo> queryPage(HouseModel model) {
    PageModel<HouseInfo> pageModel = new PageModel<>();
    String title = "%" + model.getTitle() + "%";
    String startDate = StringUtils.isNullOrEmpty(model.getStartDate()) ? "1970-01-01" : model.getStartDate();
    String endDate = StringUtils.isNullOrEmpty(model.getEndDate()) ? "2999-01-01" : model.getEndDate();
    int startIndex = (model.getCurrentPage() - 1) * model.getPageSize();
    List<HouseInfo> list = repository.pageList(title, startDate, endDate, startIndex, model.getPageSize());
    int count = repository.countPageList(title, startDate, endDate);
    pageModel.setList(list);
    pageModel.setTotalCount(count);
    return pageModel;
}
```

repository中的实现方法如下：

```
@Query(value = "select * from service_house_info where del_flag='0' " +
    " and (title like ?1 or '%%' =?1)" +
    " and (create_date >=?2 or ''=?2)" +
    " and (create_date <=?3 or ''=?3)" +
    " order by create_date desc limit ?4,?5 ", nativeQuery = true)
List<HouseInfo> pageList(String title, String startDate, String endDate, int currentPage, int pageSize);

@Query(value = "select count(1) from service_house_info where del_flag='0'" +
    " and (title like ?1 or '%%' =?1)" +
    " and (create_date >=?2 or ''=?2)" +
    " and (create_date <=?3 or ''=?3)", nativeQuery = true)
int countPageList(String title, String startDate, String endDate);
```

（2）添加或修改房源信息

```
@Override
public Result save(HouseModel model) {
    HouseInfo preInfo = new HouseInfo();
    String id = model.getId();
    if (!StringUtils.isNullOrEmpty(id)) {
        HouseInfo info = repository.findFirstById(id);
        preInfo.setCreateDate(info.getCreateDate());
    }
    BeanUtils.copyProperties(model, preInfo);
    return ResultGenerator.genSuccessResult(repository.save(preInfo));
}
```

(3）删除房源

```java
@Override
public Result delete(HouseModel model) {
    try {
        repository.delete(model.getDelFlag(), model.getId());
        return ResultGenerator.genSuccessResult("删除成功");
    } catch (Exception e) {
        e.printStackTrace();
        return ResultGenerator.genFailResult("删除失败");
    }
}
```

3.4.6 文件操作

（1）文件上传

后台接收文件类型参数，为文件生成唯一识别码，调用FileUtil将文件上传到服务器。

```java
@RequestMapping(value = "/upload", method = RequestMethod.POST)
public Result upload(@RequestParam("file") MultipartFile[] files) {
    MultipartFile file = files[0];
    log.info("图片上传: type");
    String fileName = UUIDUtils.getUUID();
    Map<String, String> map = new HashMap<>();
    String fileFullName = fileName + getFileExtension(file);
    map.put("url", "/" + getToday8Date() + "/" + fileFullName);
    map.put("file", filePath + getToday8Date() + "/" + fileFullName);
    if (FileUtil.upload(file, filePath, fileFullName)) {
        // 上传成功，给出页面提示
        log.info("图片上传结果: " + JSON.toJSONString(map));
        return ResultGenerator.genSuccessResult(map);
    } else {
        log.error("图片上传结果: " + JSON.toJSONString(map));
        throw new RuntimeException("图片上传失败");
    }
}
```

FileUtil上传文件接口，文件上传后保存到服务器指定的路径。

```java
/**
 * @param file      文件
 * @param path      文件存放路径
 * @param fileName  源文件名
 * @return
 */
public static boolean upload(MultipartFile file, String path, String fileName) {
    //先判断当前系统的顶级路径是否存在
    String realPathParent = path + File.separator + DateUtil.getToday8Date();
    File parent = new File(realPathParent);
    //判断文件父目录是否存在
    if (!parent.getParentFile().exists()) {
        parent.getParentFile().mkdir();
```

```
    }
    //使用原文件名
    String realPath = path + File.separator + DateUtil.getToday8Date() +
File.separator + fileName;
    File dest = new File(realPath);
    //判断文件父目录是否存在
    if (!dest.getParentFile().exists()) {
        dest.getParentFile().mkdir();
    }
    try {
        //保存文件
        file.transferTo(dest);
        return true;
    } catch (IllegalStateException e) {
        e.printStackTrace();
        return false;
    } catch (IOException e) {
        e.printStackTrace();
        return false;
    }
}
```

（2）文件下载

根据前端传递过来的文件路径去服务器下载对应文件，写入字节流，刷新到 HttpServletResponse中。

```
@RequestMapping(value = "/download", method = RequestMethod.POST)
public void download(@RequestBody FileModel model, HttpServletResponse response) {
    InputStream fis = null;
    OutputStream toClient = null;
    try {
        // path是指欲下载的文件的路径。
        String dirPath = model.getPath();
        String fileType = model.getFileType();
        if (StringUtils.isNotBlank(fileType)) {
            dirPath = configProperties.getFilePath() + "/" + dirPath;
        }
        // 取得文件名
        String filename = model.getFileName();
        if (StringUtils.isNotBlank(filename) && filename.equals("downDemo")) {
            dirPath = configProperties.getUserDemoFile();
        } else {
            dirPath = dirPath.replace("//", "/");
        }
        File file = new File(dirPath);
        fis = new BufferedInputStream(new FileInputStream(file));
        byte[] buffer = new byte[fis.available()];
        fis.read(buffer);
        fis.close();
        // 清空response
        response.reset();
```

```
            // 解决跨域问题,这句话是关键,对任意的域都可以,如果需要安全,可以设置成安全的域名
            response.addHeader("Access-Control-Allow-Origin","*");
            response.addHeader("Access-Control-Allow-Origin",
configProperties.getFileCros());
            response.setHeader("Access-Control-Max-Age", "3600");
            response.addHeader("Access-Control-Allow-Credentials", "true");
            response.addHeader("Access-Control-Allow-Methods", "*");
            // 设置response的Header
            response.setHeader("Content-Disposition", "attachment;filename="
                + new String(filename.getBytes(), "iso-8859-1"));
            response.setContentType("application/octet-stream");
            response.getOutputStream().write(buffer);
            response.flushBuffer();
        } catch (IOException ex) {
            ex.printStackTrace();
        } finally {
            try { // close input stream
                if (fis != null) {
                    fis.close();
                }
            } catch (Exception ex) {
                ex.printStackTrace();
            }
            try { // close output stream
                if (toClient != null) {
                    toClient.close();
                }
            } catch (Exception ex) {
                ex.printStackTrace();
            }
        }
    }
```

3.4.7 其他功能管理

除前面讲述的几个功能外,还有楼盘动态管理、认购管理、销售管理、认筹管理、楼盘装修类型管理、置业顾问管理、户型管理、物业管理、开发商管理、建筑类型管理、帮助中心、优惠券管理等功能,这些模块在功能实现上基本类似,主要涉及搜索、新增、修改、删除等功能,由于篇幅有限,不再一一详述代码。读者可参考本书配套的源码。

3.5 项目总结

本章实现了一个二手房管理系统的设计与开发,后端使用的主要技术有Spring、Spring MVC、Spring Data JPA,前端使用的技术主要有Element UI和Node.js,数据库使用的是MySQL。本章要掌握的重点是Spring Data JPA的使用,知道如何通过Spring Boot整合该框架,以及如何

在实际开发过程中使用这个框架。学习时可参考Spring官方提供的API手册，同时要与第2章所介绍的MyBatis做简单的对比，以加深理解。

本章的重点技术：

- Spring MVC框架：四种请求方式Get、Put、Post、Delete的实现方式和应用场景。
- Spring Data JPA框架：基于JPA规范实现的数据访问层框架。
- Vue：一套用于构建用户页面的渐进式框架。
- Element UI：一个基于Vue的组件库，能帮助开发者快速开发项目。
- Node.js：一个开源与跨平台的 JavaScript 运行时环境，借助它可以单独运行前端服务。
- 异步加载数据：通过Ajax实现异步获取数据并加载到页面中。

第 4 章 购物车管理系统

在第3章中，我们使用Spring Data JPA开发了二手房管理系统。本章在前一章的基础上，继续使用Spring Data JPA作为数据库接口框架开发购物车管理系统。之前的项目采用的均为MySQL数据库，本章采用的是H2数据库，页面采用的是Thymeleaf框架。

本章主要涉及的知识点有：

- 如何使用Spring Boot集成H2数据库。
- 如何使用Spring Boot集成Thymeleaf模板。
- 如何使用Spring Security安全框架。
- 如何使用H2进行数据的增、删、改、查。

4.1 项目技术选型

本章涉及的购物车管理系统功能简单，主要讲解的是所用框架技术，所以整个演示项目未采取前后端分离技术。业务代码选用的框架是Spring MVC、Spring Data JPA，安全框架采用的是Spring Security，数据库使用的是H2，页面选用的是Thymeleaf模板技术。

4.1.1 Spring Security

Spring Security是一个提供身份验证、授权和防止常见攻击的框架。由于它对命令式以及响应式程序提供的一流支持，使得它成为基于Spring的应用程序实际上的安全标准。

首先来看一下Spring Security的架构原理。Spring Security的Servlet支持是基于Servlet过滤器的，所以先了解一下过滤器的作用。图4.1的过滤器链显示了单个HTTP请求的处理程序的典型分层。

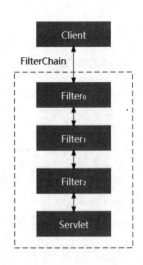

图4.1 过滤器链

客户端向应用程序发送一个请求，容器创建一条过滤器链，这条过滤器链包含多个过滤器和对应的Servlet，Servlet用来处理HttpServletRequest请求。在 Spring MVC 应用程序中，Servlet是DispatcherServlet的实例。一个Servlet最多可以处理一个HttpServletRequest和一个HttpServletResponse请求，但是，可以使用多个过滤器，这样做的目的有两种：

（1）防止下游过滤器或Servlet被调用。在这种情况下，过滤器通常会重写HttpServletResponse。

（2）修改下游过滤器或Servlet使用的HttpServletRequest和HttpServletResponse请求。

由于过滤器只影响下游的过滤器和Servlet，因此每个过滤器的调用顺序非常重要。

Spring针对这个过程提供了一个过滤器的实现类DelegatingFilterProxy，这个类实现了一种代理过滤器机制，它相当于一个容器，在这个容器内，我们可以自定义Spring Bean。这个Bean要实现过滤器接口，这样就可以将要处理的任务委派给自定义的Spring Bean，如此就既能满足过滤器链的标准，同时Spring又能够托管过滤器。

Spring Security正是利用这种过滤器委托代理机制来实现过滤的。它提供了一个FilterChainProxy类，该类就相当于上面讲的自定义Bean，通过这个FilterChainProxy可以注册多个过滤器实例，这一组实例也是一条过滤器链，叫作SecurityFilterChain，该链可以加载多个自定义的过滤器。

过滤器是根据URL地址来调用的。一条过滤器链对应的URL规则是相同的，如果要针对不同URL加载不同的过滤器链，可以创建多条过滤器链，由FilterChainProxy 决定使用哪条过滤器链。当一个请求过来后，FilterChainProxy 会按照过滤器链的顺序逐一匹配，一旦出现匹配的过滤器链，那么该过滤器链就会被调用。过滤器链相互之间都是独立、互不影响的，如图4.2所示。如果不想使用Spring Security对请求进行拦截，可以不设置过滤器。

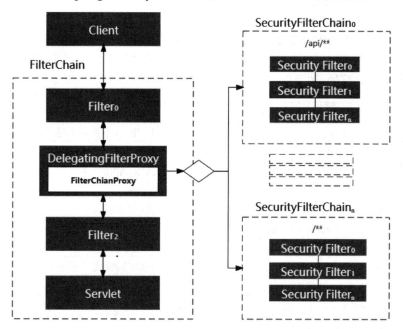

图4.2　多个SecurityFilterChain

Spring Security提供了多个过滤器，每个过滤器各司其职，如前面所说，它们有固定的加载顺序。这些过滤器的名称和顺序不需要记忆，但要明白它们的工作原理。表4.1所示是Spring Security提供的所有过滤器的完整列表。

表4.1 Spring Security过滤器列表

顺序	名称	作用
1	ChannelProcessingFilter	通道处理过滤器，判断Web请求的通道，即HTTP或HTTPS，加载不同的处理器
2	WebAsyncManagerIntegrationFilter	用于集成SecurityContext到Spring异步执行机制中的WebAsyncManager
3	SecurityContextPersistenceFilter	将有效的认证信息SecurityContext缓存到SecurityContextHolder中
4	HeaderWriterFilter	为响应对象增加一些头部信息
5	CorsFilter	跨域过滤器
6	CsrfFilter	验证是否包含系统生成的csrf的token信息
7	LogoutFilter	注销过滤器
8	OAuth2AuthorizationRequestRedirectFilter	OAuth2授权请求重定向过滤器
9	Saml2WebSsoAuthenticationRequestFilter	基于SMAL的SSO单点登录请求认证过滤器
10	X509AuthenticationFilter	认证过滤器
11	AbstractPreAuthenticatedProcessingFilter	处理经过预先认证的身份验证请求的过滤器
12	CasAuthenticationFilter	单点登录认证过滤器
13	OAuth2LoginAuthenticationFilter	OAuth2登录认证过滤器
14	Saml2WebSsoAuthenticationFilter	基于SMAL的SSO单点登录认证过滤器
15	UsernamePasswordAuthenticationFilter	处理用户以及密码认证的核心过滤器
16	OpenIDAuthenticationFilter	基于OpenID认证协议的认证过滤器
17	DefaultLoginPageGeneratingFilter	生成默认的登录页，默认/login
18	DefaultLogoutPageGeneratingFilter	生成默认的退出页，默认/logout
19	ConcurrentSessionFilter	用来判断session是否过期以及更新最新的访问时间
20	DigestAuthenticationFilter	处理HTTP头中显示的摘要式身份验证凭据
21	BearerTokenAuthenticationFilter	令牌授权过滤器
22	BasicAuthenticationFilter	处理HTTP头中显示的基本身份验证凭据
23	RequestCacheAwareFilter	用于用户认证成功后重新继续认证之前的请求
24	SecurityContextHolderAwareRequestFilter	实现Servlet API的一些接口，例如getRemoteUser、isUserInRole
25	JaasApiIntegrationFilter	Java认证授权服务
26	RememberMeAuthenticationFilter	处理"记住我"功能的过滤器
27	AnonymousAuthenticationFilter	匿名身份验证过滤器
28	OAuth2AuthorizationCodeGrantFilter	OAuth2授权码模式过滤器
29	SessionManagementFilter	会话管理过滤器
30	ExceptionTranslationFilter	传输异常事件
31	FilterSecurityInterceptor	动态权限控制

通过过滤器的请求会进入安全异常处理阶段，如果通过了异常处理即未抛出AccessDeniedException，则进入身份认证阶段，否则调用AccessDeniedHandler处理器处理异常信息，如图4.3所示。

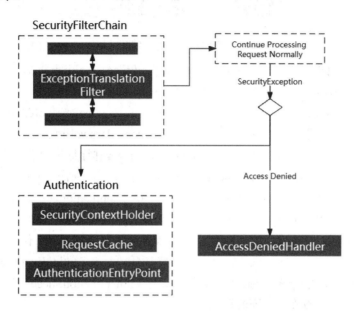

图4.3 安全异常处理

上述流程的实现如下：

首先，ExceptionTranslationFilter调用FilterChain.doFilter(request, response)来调用应用程序的其余部分。如果用户未通过身份验证或它是一个AuthenticationException，则开始身份验证。验证过程如下：

步骤01 首先清空SecurityContextHolder中的内容。

步骤02 HttpServletRequest保存在RequestCache。当用户成功认证后，RequestCache用于重放原始请求。

步骤03 AuthenticationEntryPoint用于从客户端请求身份验证凭据。例如，它可能会重定向到登录页面或发送WWW-Authenticate标头。

如果是AccessDeniedException，则拒绝访问，调用AccessDeniedHandler来处理拒绝访问。Spring Security身份验证模型的核心是SecurityContextHolder。经过SecurityContextPersistenceFilter过滤器的处理后，身份验证者的详细信息会被存储在SecurityContextHolder中。SecurityContextHolder中包含SecurityContext，SecurityContext中包含一个Authentication对象。Authentication对象可以为身份认证管理器提供用户的身份验证凭据，也可以表示当前经过身份验证的用户。一个Authentication包含Principal（用户信息）、Credentials（密码）和Authorities（角色或权限）三部分，如图4.4所示。

用户名和密码过滤器认证时，需要调用认证器。认证器由身份验证管理器AuthenticationManager提供，它是定义Spring Security的过滤器如何执行身份验证的API。在Spring Security中，最常用的AuthenticationManager实现是ProviderManager。

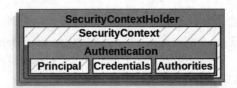

图4.4　SecurityContextHolder的组成

当用户提交他们的凭据时，AbstractAuthenticationProcessingFilter将基于HttpServletRequest创建一个Authentication进行身份验证。Authentication的类型取决于AbstractAuthenticationProcessingFilter的子类。例如，UsernamePasswordAuthenticationFilter用HttpServletRequest提交的用户名和密码创建一个UsernamePasswordAuthenticationToken。

接下来，将Authentication传递给AuthenticationManager进行身份验证。

如果身份验证失败，则SecurityContextHolder被清除；调用RememberMeServices.loginFail，如果没有配置"记住我"，则跳过该操作；然后调用AuthenticationFailureHandler。

如果身份验证成功，则SessionAuthenticationStrategy收到新的登录通知，Authentication在SecurityContextHolder上进行设置。然后SecurityContextPersistenceFilter保存SecurityContext到HttpSession。RememberMeServices.loginSuccess被调用。ApplicationEventPublisher发布一个InteractiveAuthenticationSuccessEvent事件，然后调用AuthenticationSuccessHandler。

以上架构分析对应的整个认证流程如图4.5所示。

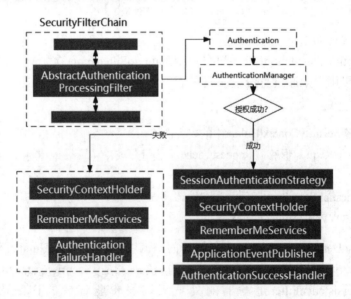

图4.5　认证流程图

4.1.2　H2数据库

1. 概述

H2是开源的轻量级嵌入式数据库引擎，是采用Java实现的关系数据库，它可以嵌入Java应用程序中或以客户端服务器模式运行。H2的官网地址为http://www.h2database.com/，在这里可以查看它的详细介绍和详细的官方API文档。类似的轻量级数据库实际上有很多，例如Java自带的Derby，还有历史悠久的HSQLDB、PostgreSQL、Sqlite等。

H2数据库作为内存数据库运行，这意味着数据将不会持久存储在磁盘上。由于H2全部由Java编写，所以可以用Java对它进行全面的控制，例如启动/停止服务、管理账号等。

H2数据库的主要特点如下：

（1）占用空间小

JAR包文件只有2.5MB，占用空间非常小，并且不依赖任何第三方组件。

（2）速度非常快

由于数据存储在内存中，所以数据的读写速度非常快，适合高频率读写的数据，例如本章涉及的购物车。

（3）操作便捷

允许用户通过浏览器接口的方式访问SQL数据库。安装完成后，启动服务不需要其他客户端软件，直接用浏览器即可访问数据库接口。

（4）开源

支持开源协议，这使得开发者能够更加清楚地知道它的运行原理，更透彻地理解它的优势和弊端。

（5）支持标准的SQL和JDBC API

H2支持标准的SQL，且支持JDBC API，这使得它的兼容性更强。

（6）支持多种连接模式

H2支持三种连接模式，嵌入式、客户端服务器连接、混合模式连接。其中嵌入式是将H2嵌入Java程序中，作为程序的一部分，这样部署和实施时非常方便，开发者不用再考虑服务器的环境问题。

（7）开发文档全面且详细

H2提供的官方文档非常全面且详细，一般使用中的问题都可以在文档中找到答案。

2. 下载和安装

步骤01 打开H2的官网，根据自己的操作系统选择对应的版本，可以选择安装文件，也可以选择zip安装包下载。这里以Windows为例，下载zip安装包，如图4.6所示。

步骤02 下载完成后打开压缩包，得到如图4.7所示目录结构。

步骤03 打开cmd控制台，进入上述解压目录的bin目录下，执行如下命令行：

```
java -cp h2-2.1.210.jar org.h2.tools.Shell
```

步骤04 根据提示依次输入URL、Driver、User和Password。数据库名称为test，连接账号为hazel，连接密码为123456，这些参数都可以按需指定，本次指定的账号将自动视为数据库test的管理员账号。执行结果如图4.8所示。

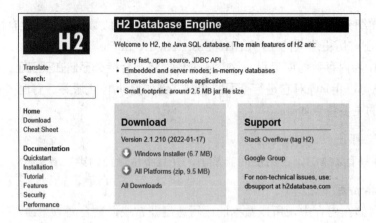

图4.6　H2下载页面

名称	修改日期	类型	大小
bin	2022/1/25 14:54	文件夹	
docs	2022/1/25 14:54	文件夹	
service	2022/1/25 14:54	文件夹	
src	2022/1/25 14:54	文件夹	
build.bat	2022/1/17 10:12	Windows 批处理...	1 KB
build.iml	2022/1/17 10:12	IML 文件	1 KB
build.sh	2022/1/17 10:12	Shell Script	1 KB

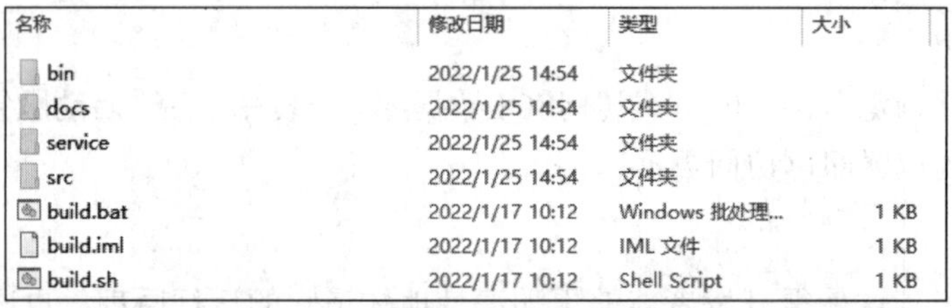

图4.7　H2目录结构

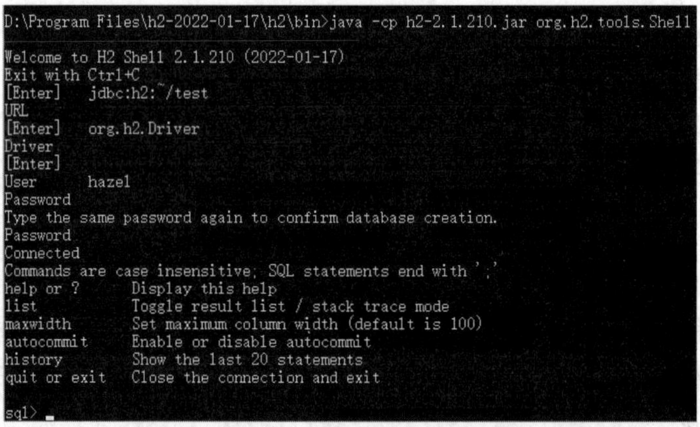

图4.8　初始化H2数据库

上述操作完成后，在项目中就可以使用JDBC连接数据库了。以上配置完成后，按快捷键Ctrl+C退出配置。

3. 启动

（1）*以Web Server方式启动*

在cmd控制台中输入如下命令以Web Server方式启动H2：

```
java -jar h2-1.4.200.jar
```

执行结果如图4.9所示。只有通过这种方式启动，H2才允许通过浏览器访问数据库，同时在系统托盘处显示对应图标，如图4.10所示。

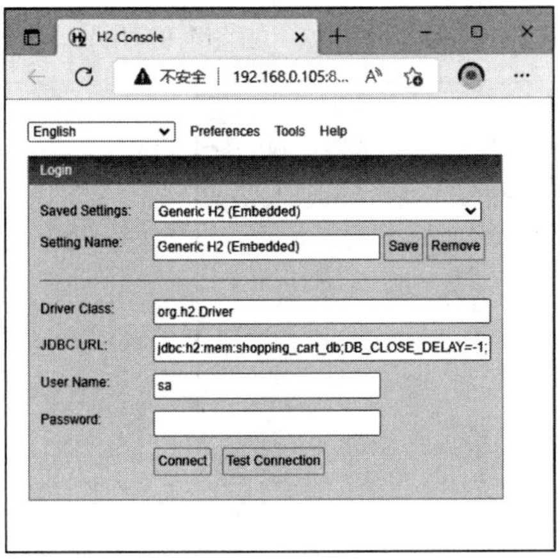

图4.9　执行结果

图4.10　H2服务系统图标

启动成功后自动弹出访问网页，H2默认端口号为8082。可以在网页上方的语言下拉框中将语言类型切换到中文，如图4.11所示。

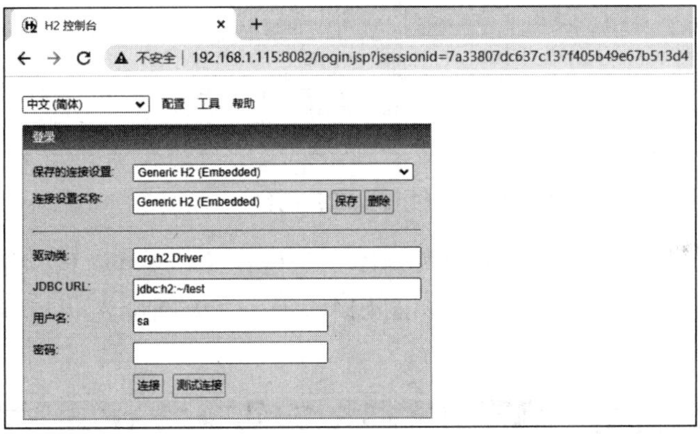

图4.11　连接H2

然后在上述页面中输入初始化配置的用户名和密码，进行连接测试，如图4.12所示。

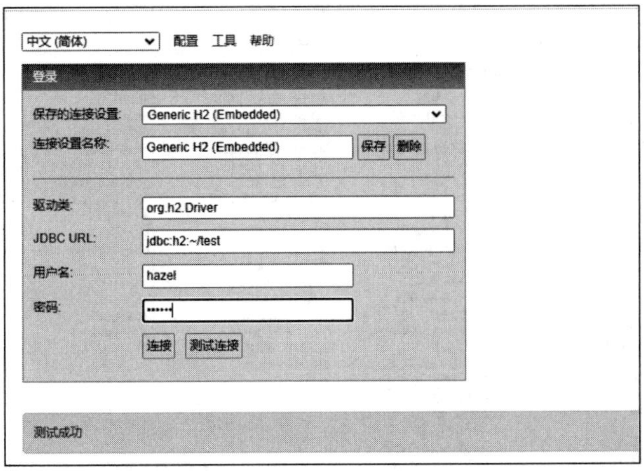

图4.12　H2数据库连接测试

测试完成后，连接H2数据库。连接成功后进入H2数据库控制台，如图4.13所示。

图4.13　H2控制台

在这个窗口中，可以使用H2的数据库脚本语言进行各项操作。H2支持标准的SQL语言规范，对应的官网也有数据库命令说明，如图4.14所示，官方网址为http://www.h2database.com/html/commands.html。

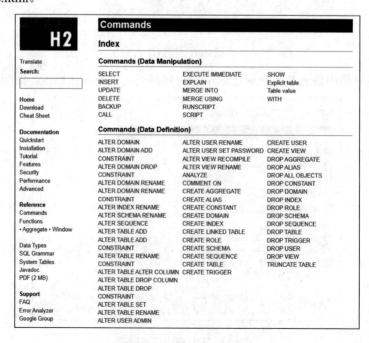

图4.14　H2支持的SQL语法

例如，执行select查询users表中的数据，H2查询页面如图4.15所示。

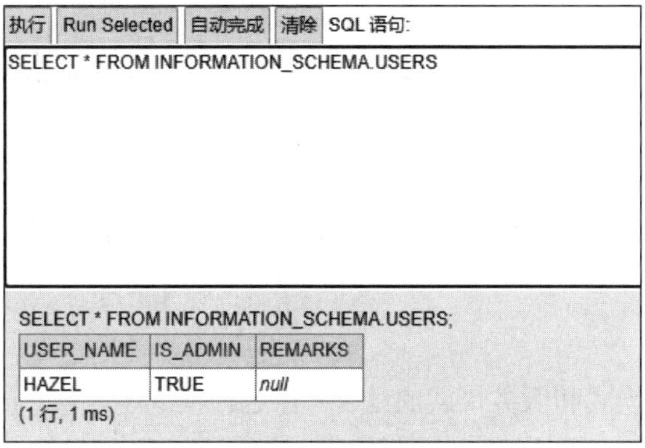

图4.15　H2查询页面

（2）以TCP Server方式启动

在命令窗口中输入如下命令：

```
java -cp h2-2.1.210.jar org.h2.tools.Server
```

如果刚才的Web Server方式仍在运行的话，可以先按快捷键Ctrl+C，退出当前运行方式，也可以输入Exit退出，退出后系统托盘的图标会消失，然后再执行上面的命令。

执行结果如图4.16所示，启动成功后同样会弹出如图4.11所示的浏览器页面。只是以这种方式启动，系统右下角没有托盘图标。

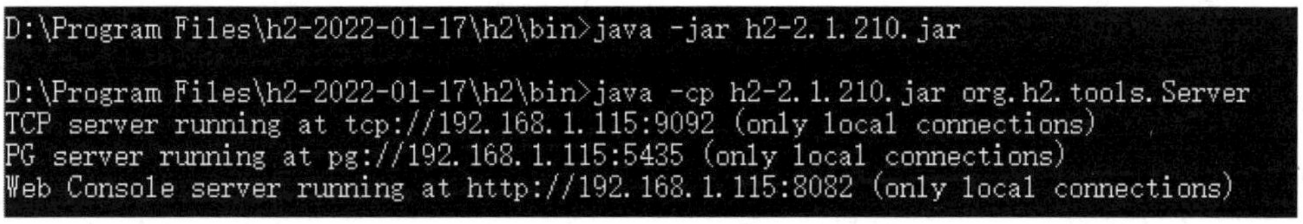

图4.16　启动H2

退出时，可以直接按快捷键Ctrl+C，也可以执行如下命令：

```
java -cp h2-2.1.210.jar org.h2.tools.Server -tcpShutdown tcp://192.168.1.115:8082 -tcpPassword 123456
```

4. 三种连接方式

连接H2数据库有3种方式：

- 嵌入模式：支持使用本地JDBC方式连接。
- 服务模式：支持通过TCP/IP协议使用远程JDBC或ODBC模式连接。
- 混合模式：支持通过本地或远程两种方式连接。

（1）嵌入模式

在嵌入模式中，数据库与应用必须在同一个JVM中，通过JDBC来连接H2数据库，但是只支持一个客户端连接。这是最便捷的连接方式，启动应用的同时，会把H2数据服务也启动，应用中同时包含了H2数据库的服务端和客户端，如图4.17所示。

本章的实战项目购物车管理系统使用的就是这种模式。

（2）服务模式

服务模式，也称为远程连接模式或客户端服务器模式。在这种模式中，通过使用JDBC或ODBC API远程连接数据库，支持多个应用同时连接同一个数据库服务端，如图4.18所示。相比于嵌入式模式，该模式在性能上会有所降低，因为所有的数据传输需要通过TCP/IP协议进行。

（3）混合模式

混合模式是嵌入式模式和服务模式的结合。在这种模式中，有一个应用以嵌入式方式连接H2数据库，与此同时，由它所启动的H2数据库服务端供其他应用连接，此时其他应用均为H2的客户端，如图4.19所示。由于服务端内嵌到某个应用中，所以可以通过该应用来启动或者停止H2服务端。

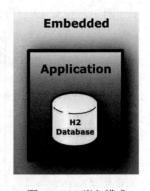

图4.17　H2嵌入模式

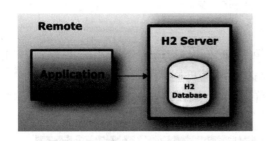

图4.18　H2服务模式

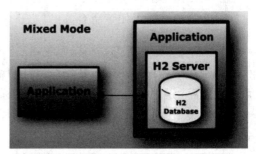

图4.19　H2混合模式

（4）数据库连接URL

由于支持多种连接模式并且有多种应用场景，所以H2数据库连接的URL会有所不同，表4.2所示是各种连接情况下的H2数据库URL规则，URL中不区分大小写。

表4.2　H2数据库连接场景URL示例

连接方式或应用场景	URL格式和示例
嵌入模式	jdbc:h2:[file:][<path>]<databaseName> jdbc:h2:~/test jdbc:h2:file:/data/sample jdbc:h2:file:C:/data/sample（仅支持Windows）
内存存储（私有空间）	jdbc:h2:mem:
内存存储（命名空间）	jdbc:h2:mem:<databaseName> jdbc:h2:mem:test_mem
远程服务端模式（TCP/IP）	jdbc:h2:tcp://<server>[:<port>]/[<path>]<databaseName> jdbc:h2:tcp://localhost/~/test jdbc:h2:tcp://dbserv:8084/~/sample jdbc:h2:tcp://localhost/mem:test
远程服务端模式（TLS）	jdbc:h2:ssl://<server>[:<port>]/[<path>]<databaseName> jdbc:h2:ssl://localhost:8085/~/sample;

（续表）

连接方式或应用场景	URL格式和示例
使用加密文件	jdbc:h2:<url>;CIPHER=AES
	jdbc:h2:ssl://localhost/~/test;CIPHER=AES
	jdbc:h2:file:~/secure;CIPHER=AES
文件锁定方法	jdbc:h2:<url>;FILE_LOCK={FILE|SOCKET|FS|NO}
	jdbc:h2:file:~/private;CIPHER=AES;FILE_LOCK=SOCKET
存在时才可打开	jdbc:h2:<url>;IFEXISTS=TRUE
	jdbc:h2:file:~/sample;IFEXISTS=TRUE
VM运行时不关闭数据库	jdbc:h2:<url>;DB_CLOSE_ON_EXIT=FALSE
连接时执行SQL	jdbc:h2:<url>;INIT=RUNSCRIPT FROM '~/create.sql'
	jdbc:h2:file:~/sample;INIT=RUNSCRIPT FROM '~/create.sql'\\;RUNSCRIPT FROM '~/populate.sql'
使用用户名和密码	jdbc:h2:<url>[;USER=<username>][;PASSWORD=<value>]
	jdbc:h2:file:~/sample;USER=sa;PASSWORD=123
Debug追踪设置	jdbc:h2:<url>;TRACE_LEVEL_FILE=<level 0..3>
	jdbc:h2:file:~/sample;TRACE_LEVEL_FILE=3
忽略未知配置	jdbc:h2:<url>;IGNORE_UNKNOWN_SETTINGS=TRUE
客户端文件接入方式	jdbc:h2:<url>;ACCESS_MODE_DATA=rws
ZIP格式的数据库文件	jdbc:h2:zip:<zipFileName>!/<databaseName>
	jdbc:h2:zip:~/db.zip!/test
兼容模式	jdbc:h2:<url>;MODE=<databaseType>
	jdbc:h2:~/test;MODE=MYSQL;DATABASE_TO_LOWER=TRUE
自动重连	jdbc:h2:<url>;AUTO_RECONNECT=TRUE
	jdbc:h2:tcp://localhost/~/test;AUTO_RECONNECT=TRUE
自动混合模式	jdbc:h2:<url>;AUTO_SERVER=TRUE
	jdbc:h2:~/test;AUTO_SERVER=TRUE
页面大小	jdbc:h2:<url>;PAGE_SIZE=512
修改其他配置	jdbc:h2:<url>;<setting>=<value>[;<setting>=<value>...]
	jdbc:h2:file:~/sample;TRACE_LEVEL_SYSTEM_OUT=3

通过Java代码使用以上URL连接H2的示例代码如下：

```
Connection conn = DriverManager.getConnection("jdbc:h2:~/test", "username",
"password");
    // ...
    conn.close();
```

在后续的项目代码中也有类似的操作语句。

4.1.3 Thymeleaf

Thymeleaf是一种时新的Java模板引擎，它在服务端解析，支持Web应用或者独立部署的单机应用。Thymeleaf能够更准确更优雅地渲染HTML。在Thymeleaf模板中出现的HTML语句，依旧能像HTML一样正常解析与渲染。除了Thymeleaf，还有一些其他的模板引擎，例如JSP、

Velocity、FreeMaker等。相比而言，Thymeleaf可以在不启动Web应用的情况下直接在浏览器中显示模板内容。另外，它能非常便捷地与Spring框架集成，对开发者来说，这点非常友好，并且能提高开发效率。

Thymeleaf遵循开箱即用原则，可以处理标记模板模式（HTML和XML）、文本模板模式（TEXT、JAVASCRIPT、CSS）和无操作模板模式（RAW）。

在HTML模板中，Thymeleaf允许任何类型的HTML输入，包括HTML5、HTML4和XHTML，不进行任何验证或格式良好性检查，并且在输出中尽可能尊重模板代码的结构。

XML模板模式允许XML输入。在这种情况下，代码应该是格式良好的，没有未闭合的标签、没有未引用的属性等，如果发现格式违规，解析器将抛出异常。

TEXT模板模式将允许对非标记性质的模板使用特殊语法。此类模板的示例可能是文本电子邮件或模板文档。注意，HTML或XML模板也可以处理为TEXT，在这种情况下，它们不会被解析为标记，并且每个标签、DOCTYPE、注释等都将被视为纯文本。

JAVASCRIPT模板模式允许在Thymeleaf应用程序中处理JavaScript文件，这意味着能够以与HTML文件中相同的方式使用JavaScript文件中的数据。

CSS模板模式允许处理Thymeleaf应用程序中涉及的 CSS 文件。与JAVASCRIPT模式类似，CSS模板模式也是一种文本模式，使用TEXT模板模式的特殊处理语法。

RAW模板模式根本不会处理模板，它用于将未触及的资源（文件、URL响应等）插入正在处理的模板中。例如，外部的HTML格式的资源可以包含在应用程序模板中，并且这些资源可能包含的任何Thymeleaf 代码都不会被执行。

Thymeleaf的核心库提供了一种称为标准方言的方言，这个方言支持自定义扩展。在实际应用过程中，Thymeleaf提供的内容已足够我们使用，无须扩展。

> **注意：** 官方的thymeleaf-spring3 和 thymeleaf-spring4 集成包都定义了一种称为SpringStandard Dialect的方言，它与标准方言大体相同，但它更好地利用了Spring Framework中的某些特性（例如，使用Spring表达式语言或SpringEL而不是OGNL）。

和任何一种工具或语言一样，Thymeleaf有自己的语法规则。接下来我们看一下常用的语法规则。

（1）引入Thymeleaf

修改html标签用于引入thymeleaf引擎，这样才可以在其他标签里使用th:*语法，这是使用Thymeleaf的前提。

```
<!DOCTYPE html SYSTEM
"http://www.thymeleaf.org/dtd/xhtml1-strict-thymeleaf-spring4-4.dtd">
    <html xmlns="http://www.w3.org/1999/xhtml" xmlns:th="http://www.thymeleaf.org">
```

xmlns属性是用来定义xml namespace（命名空间）的。该属性可以放置在文档内任何元素的开始标签中。该属性的值类似于URL，它定义了一个命名空间，浏览器会将此命名空间用于该属性所在元素内的所有内容。引入之后就可以使用th作为前缀使用Thymeleaf相关的标签了。严格来说，这并不是HTML5的规范写法，如果要使整个模板完全符合HTML规范，那么我们可以不使用命名空间引入的方法，而将th:改写为data-th-作为属性前缀。

```
<p data-th-text="#{home.welcome}">欢迎光临我们的商城!</p>
```

(2) 外部化文本

外部化文本是从模板文件中提取代码对应的内容,将它们保存在单独的文件(通常是.properties文件)中,并且可以很容易地用其他语言编写的等效文本替换(称为国际化或简称为i18n的过程)。这种外部化文本通常称为"消息"。

消息总是有一个标识它们的键,Thymeleaf允许使用#{...}语法指定文本对应的特定消息:

```
<p th:text="#{home.welcome}">欢迎光临我们的商城!</p>
```

在这里实际上是Thymeleaf标准方言的两个特征:

- th:text属性用动态取得的表达式内容替换掉标签中设置的默认值。
- 标准表达式语法#{home.welcome}指定了该表达式对应的是home.welcome这个key对应的消息内容。

这个外部化的文本在哪里?

Thymeleaf中外部化文本的位置是可配置的,这将取决于org.thymeleaf.messageresolver. IMessageResolver所使用的具体实现。通常,将使用基于.properties文件的实现,但如果想要从数据库获取消息,我们可以创建自己的实现。

对于Spring Boot来说,Spring Boot为我们提供了国际化的配置。文本文件的前缀默认为message,例如message.properties、message_en_US.properties、message_zh_CN.properties。默认情况下,Spring Boot会自动加载这些文件。

```
/templates/home_zh_CN.properties对于中文文本
/templates/home_en_US.properties对于英文文本
```

这个message前缀可以在Spring Boot配置文件中配置。

```
spring
  messages:
    basename: i18n.login  #默认为i18n.messages
```

接下来只需要在resources/i18n路径下创建相应的文件,文件开头为上述已配置的前缀,例如login.properties、login_en_US.properties、login_zh_CN.properties。

(3) 标准表达式语法

- 简单的表达
    - 变量表达式: ${...}。
    - 选择变量表达式: *{...}。
    - 消息表达式: #{...}。
    - 链接URL表达式: @{...}。
    - 片段表达式: ~{...}。
- 字面量
    - 文本字面量: 'ABC', 'Hello!',......
    - 数字文字: 0, 34, 3.0, 12.3,......

- 布尔文字：true,false。
- 空文字：null。
- 文字标记：one, sometext, main,......
* 文字操作
  - 字符串连接：+。
  - 字面替换：|The name is ${name}|。
* 算术运算
  - 二元运算符：+，—，*，/，%。
  - 减号（一元运算符）：—。
* 布尔运算
  - 二元运算符：and,or。
  - 布尔否定（一元运算符）：!：,not。
* 比较和平等
  - 比较器：>, <, >=, <=(gt, lt, ge, le)
  - 等式运算符：==, !=(eq, ne)
* 条件运算符
  - 如果—那么：(if) ? (then)
  - 如果—那么—否则：(if) ? (then) : (else)
  - 默认：(value) ?: (defaultvalue)
* 特殊代币
  - 无操作：_

所有这些功能都可以组合和嵌套，例如：

```
'用户类型 ' + (${user.isAdmin()} ? 'Administrator' : (${user.type} ?: 'Unknown'))
```

（4）变量

这里适用的变量既可以是基本类型也可以是对象。Thymeleaf通过${}来获取model中的变量，但这不是EL表达式，而是OGNL表达式，两者语法非常相像。

例如，Student对象中有姓名stuName属性，那么我们可以取出对应的姓名展示出来，代码如下：

```
<span th:text="${student.stuName}">江小白</span>
<input type="text" name="stuName" value="江小白" th:value="${student.stuName}" />
```

在不启动项目的情况下，该页面也可以打开，只是无法取出实际的姓名参数值，而是展示默认值"江小白"。这就是Thymeleaf的优雅之处，减少了设计和开发团队之间的沟通障碍。

上述写法是HTML5的写法，如果浏览器不支持HTML5，那么可以考虑将th:text写成data-th-text。这种写法依然可以被Thymeleaf解析。

与th:text标签对应的还有一个th:utext标签，这两个标签基本相同，区别在于th:text输出的内容是经过转义之后的，而th:utext输出的为原始内容。

如果一个变量有多个属性，可以采用选择表达式，即*写法，用*代替变量，省去变量名。

```html
<div th:object="${user}">
    <p>姓名：<span th:text="*{name}">江小白</span></p>
    <p>性别：<span th:text="*{sex}">男</span></p>
    <p>年龄：<span th:text="*{age}">20</span></p>
    <p>居住地：<span th:text="*{address}">上海</span></p>
</div>
```

看到这里会发现，Thymeleaf的表达式有3种符号，"$""#"和"*"。

$符号取上下文中的变量：

```html
<input type="text" name="userName" th:value="${user.name}">
```

#符号取Thymeleaf内置的方法、文字消息表达式：

```html
<p th:utext="#{home.welcome}">Welcome to our grocery store!</p>
<span th:text="${#calendars.format(today,'dd MMMM yyyy')}">13 May 2011</span>
```

*{...}选择表达式一般跟在th:object后，直接选择object中的属性。

注意：在启用 Spring MVC 的应用程序中，OGNL 将替换为SpringEL，但其语法与 OGNL 非常相似（实际上，对于大多数常见情况两者完全相同）。

（5）ONGL表达式基本对象

在上下文变量上评估OGNL表达式时，某些对象可用于表达式以获得更高的灵活性。#对象将被引用（根据OGNL标准），以符号开头：

- #ctx：上下文对象。
- #vars：上下文变量。
- #locale：上下文语言环境。
- #request：（仅在 Web 上下文中）HttpServletRequest对象。
- #response：（仅在 Web 上下文中）HttpServletResponse对象。
- #session：（仅在 Web 上下文中）HttpSession对象。
- #servletContext：（仅在 Web 上下文中）ServletContext对象。

所以我们可以这样做：

```html
Established locale country: <span th:text="${#locale.country}">US</span>
```

（6）表达式实用程序对象

除了这些基本对象之外，Thymeleaf还提供一组实用对象，这些对象将帮助我们在表达式中执行常见任务。

- #execInfo：有关正在处理的模板的信息。
- #messages：在变量表达式中获取外部化消息的方法，与使用 #{...} 语法获取它们的方式相同。
- #uris：转义部分 URL/URI 的方法。
- #conversions：执行配置的转换服务的方法（如果有的话）。
- #dates：java.util.Date对象的方法，格式化、组件提取等。
- #calendars：类似于#dates，但用于java.util.Calendar对象。

- #numbers：格式化数字对象的方法。
- #strings：包含String对象的方法，例如startsWith、endsWith、indexOf、length等。
- #objects：一般对象的方法。
- #bools：布尔评估方法。
- #arrays：数组的方法。
- #lists：列表的方法。
- #sets：集合的方法。
- #maps：地图的方法。
- #aggregates：在数组或集合上创建聚合的方法。
- #ids：处理可能重复的id属性的方法（例如，作为迭代的结果）。

（7）在页面格式化日期

ONGL表达式中内置了许多方法，例如字符串的拆分与处理、日期格式化等。使用时可以直接使用变量调用该方法。

```
<p>
   Today is: <span th:text="${#calendars.format(today,'dd MMMM yyyy')}">13 May 2011</span></p>
```

（8）链接URL

使用url标签时，需要使用@符号。

```
<a href="/user/info"  th:href="@{/user/info(id=${user.id})}" >查看信息</a>
```

注意：th:href是修饰符属性，它将计算要使用的链接URL并将该值设置为<a>标签的href属性。

可以对URL参数使用表达式，如id=${user.id}，所需的URL参数编码操作也将自动执行。如果需要多个参数，这些参数将用逗号分隔，如@{/user/query(name=${name},age=18)} URL路径中也允许使用变量模板，如@{/user/{userId}/details(userId=${user.id})}

以 "/" 开头的相对路径将自动以应用程序上下文名称为前缀。

如果cookie未启用或未知，则";jsessionid=XXX"可能会在相对 URL 中添加后缀，以便保留会话，这称为URL重写。Thymeleaf 允许使用response.encodeURL(...)为 URL插入自己的重写过滤器。

该th:href属性允许在模板中设置一个静态href属性，以便模板链接在未启动Spring项目直接打开时仍然可以由浏览器导航。

（9）片段

片段表达式可以标记片段，并在模板周围移动片段。片段可以被复制，也可以将它们作为参数传递给其他模板，等等。最常见的用途是使用th:insert或者th:replaceor进行片段插入或替换。

```
<div th:insert="~{commons :: main}">...</div>
<div th:with="frag=~{footer :: #main/text()}">
<p th:insert="${frag}"></div>
```

（10）运算符和比较器

可以使用一些算术运算：+、–、*、/和%。这些算术运算符写在大括号之外的话，由Thymeleaf负责处理。

```
<div th:with="isEven=(${prodStat.count} % 2 == 0)">
```

写在大括号之内的话，由OGNL或SpringEL负责处理。

```
<div th:with="isEven=${prodStat.count % 2 == 0}">
```

表达式中的值可以使用>、<、>=和<=符号进行比较，==和!=运算符可用于检查是否相等（或不相等）。注意，XML 规定不应在属性值中使用<或>符号，因此应将它们替换为<或>。

```
<div th:if="${prodStat.count} &gt; 1"><span th:text="'Execution mode is ' +
( (${execMode} == 'dev')? 'Development' : 'Production')">
```

（11）条件表达式

条件表达式类似三目表达式。

```
<tr th:class="${row.even}? 'even' : 'odd'">
    ...</tr>
```

支持嵌套使用。

```
<tr th:class="${row.even}? (${row.first}? 'first' : 'even') : 'odd'">
    ...</tr>
```

（12）遍历

遍历所使用的命令为th:each。例如，查询到许多学生信息，将它们放入一个list中，变量名为students，在表格中每一行展示一个学生的信息，就可以用下面的方式遍历所有学生的信息。

```
<tr th:each="stu: ${students}">
    <td th:text="${stu.name}">Jiang Xiaobai</td>
    <td th:text="${stu.age}">21</td>
    <td th:text="${stu.sex == 1}? '男' : '女'">未知</td>
</tr>
```

可迭代的值不只有list，Thymeleaf支持很多列表形式的数据迭代。

- 任何实现的对象java.util.Iterable。
- 任何实现java.util.Enumeration。
- 任何实现的对象java.util.Iterator，其值将在迭代器返回时使用，而无须将所有值缓存在内存中。
- 任何实现java.util.Map，迭代映射时，迭代变量将属于 class java.util.Map.Entry。
- 任何数组。
- 任何其他对象都将被视为包含对象本身的单值列表。

4.1.4　框架搭建

本章项目选取的是购物车管理系统，以下为该项目的技术框架部分：

- 开发框架：Spring Boot。
- 数据库：H2。
- 后台框架：Spring、Spring MVC、Spring Data JPA、Spring Security。
- 前端框架：Thymeleaf。

Spring、Spring MVC以及Spring Data JPA的整合在第3章中已经介绍过了，在本章中继续沿用。在页面部分，采用页面模板解析技术Thymeleaf。

在Spring Boot中集成H2和Thymeleaf非常简单，只需引入对应依赖即可。相关依赖如下所示。

```xml
<!--H2数据库-->
<dependency>
    <groupId>com.h2database</groupId>
    <artifactId>h2</artifactId>
    <scope>runtime</scope>
</dependency>

<!--Thymeleaf相关依赖-->
<dependency>
    <groupId>org.springframework.boot</groupId>
    <artifactId>spring-boot-starter-thymeleaf</artifactId>
</dependency>
<dependency>
    <groupId>org.thymeleaf.extras</groupId>
    <artifactId>thymeleaf-extras-springsecurity5</artifactId>
</dependency>
```

引入完成后，需要在Spring Boot配置文件中配置对应参数。首先配置H2相关信息，代码如下：

```
spring.datasource.url=jdbc:h2:mem:shopping_cart_db;DB_CLOSE_DELAY=-1;DB_CLOSE_ON_EXIT=FALSE
spring.datasource.username=sa
spring.datasource.password=
spring.sql.init.data-locations=classpath:/sql/import-h2.sql
spring.main.allow-bean-definition-overriding=true
spring.jpa.show-sql=true
spring.jpa.defer-datasource-initialization=true
spring.h2.console.enabled=true
spring.h2.console.path=/h2-console
```

url为数据库链接地址，具体规则可以参考前面H2介绍部分。这里DataSource中的data为数据源，因为使用的是H2的内嵌模式，所以可以在项目启动时通过SQL语句和该项配置将初始数据导入数据库中。spring.jpa.defer-datasource-initialization设置为true，代表允许项目启动时通过实体类初始化数据库结构。Console为控制台相关配置，在开发过程中使用。

然后配置Thymeleaf。这里只配置了模板存放的目录以及缓存是否开启。

```
# Thymeleaf
spring.thymeleaf.cache=false
spring.thymeleaf.prefix=classpath:/templates
```

接下来配置Spring Security。首先引入相关依赖，如下所示。

```xml
<!--安全框架-->
<dependency>
    <groupId>org.springframework.boot</groupId>
    <artifactId>spring-boot-starter-security</artifactId>
</dependency>
```

最后配置相关信息。首先在Spring配置文件中配置用户权限相关的查询语句。

```
# Spring Security
# Queries for AuthenticationManagerBuilder
spring.queries.users-query=select username, password, active from t_user where username=?
spring.queries.roles-query=select u.username, r.role from t_user u inner join user_role ur on(u.user_id=ur.user_id) inner join t_role r on(ur.role_id=r.role_id) where u.username=?
```

其中spring.queries.users-query配置根据用户名查询用户信息的语句，spring.queries.roles-query配置根据用户名查询角色信息的语句。完成以上配置后，编写自定义SecurityConfig类。在5.7版本以前，需要继承自Spring Security提供的适配器WebSecurityConfigurerAdapter。5.7版本之后，直接自定义即可。本章项目采用的Spring Boot版本为2.7.1，对应的Spring Security版本在5.7之后，所以该类的具体代码如下：

```java
@Configuration
@RequiredArgsConstructor
public class SecurityConfig{
    private final DataSource dataSource;//数据源
    @Value("${spring.queries.users-query}")
    private String usersQuery;//用户信息查询语句
    @Value("${spring.queries.roles-query}")
    private String rolesQuery;//角色信息查询语句
    private final CustomAccessDeniedHandler accessDeniedHandler;//未通过安全校验的处理器
    /**
     * 用户与权限校验查询语句
     * @return
     */
    @Bean
    UserDetailsService userDetailsService() {
        //针对用户数据存放于数据库的情况，使用JdbcUserDetailsManager类进行查询与校验
        JdbcUserDetailsManager users = new JdbcUserDetailsManager();
        //设置数据源
        users.setDataSource(dataSource);
        //查询用户语句
        users.setUsersByUsernameQuery(usersQuery);
        //查询权限语句
        users.setAuthoritiesByUsernameQuery(rolesQuery);
        return users;
    }
    /**
```

```java
     * 过滤器链
     * @param http
     * @return
     * @throws Exception
     */
    @Bean
    SecurityFilterChain securityFilterChain(HttpSecurity http) throws Exception {
        http.csrf().disable()
            .authorizeRequests()
            //不需要拦截的URL规则
            .antMatchers("/", "/home", "/registration", "/error", "/h2-console/**"
                    , "/css/**", "/js/**", "/images/**", "/webjars/**").permitAll()
            .anyRequest().authenticated()
            .and()
            .formLogin()
            //登录页
            .loginPage("/login")
            //默认成功返回URL
            .defaultSuccessUrl("/home")
            //登录页和默认页不需要拦截
            .permitAll()
            .and()
            .logout()
            .permitAll()
            .and()
            //设置用户相关的数据库操作
            .userDetailsService(userDetailsService())
            //配置未成功通过安全认证的处理器
            .exceptionHandling().accessDeniedHandler(accessDeniedHandler)
            // 作用于h2-console
            .and().headers().frameOptions().disable();
        return http.build();
    }
    /**
     * 初始化认证管理器
     * @param authenticationConfiguration
     * @return
     * @throws Exception
     */
    @Bean
    public AuthenticationManager authenticationManager
(AuthenticationConfiguration authenticationConfiguration) throws Exception{
        AuthenticationManager authenticationManager = authenticationConfiguration.getAuthenticationManager();
        return authenticationManager;
    }
    /**
     * 密码加密方式
     * 默认返回BCryptPasswordEncoder的PasswordEncoder
     */
    @Bean
```

```
    public PasswordEncoder passwordEncoder() {
        return new BCryptPasswordEncoder();
    }
}
```

类声明语句上添加了两个注解@Configuration和@RequiredArgsConstructor。@Configuration注解代表该类是一个配置类。@RequiredArgsConstructor是工具lombok提供的一个注解，与前面用到的@Data注解属于同一个工具包。使用时必须导入lombok相关依赖。

```
<dependency>
    <groupId>org.projectlombok</groupId>
    <artifactId>lombok</artifactId>
    <version>1.18.12</version>
    <scope>provided</scope>
</dependency>
```

@RequiredArgsConstructor用来生成带有参数的构造函数，所需参数必须以final声明，例如SecurityConfig类中的CustomAccessDeniedHandler和DataSource。

进行拦截配置时，需要两个查询语句usersQuery和rolesQuery，这个可以通过@Value注解从Spring配置文件中获取。对应的配置如下：

```
# Spring Security
# Queries for AuthenticationManagerBuilder
spring.queries.users-query=select username, password, active from t_user where username=?
spring.queries.roles-query=select u.username, r.role from t_user u inner join user_role ur on(u.user_id=ur.user_id) inner join t_role r on(ur.role_id=r.role_id) where u.username=?
```

在5.7版本之前，SecurityConfig的核心方法是configure方法。在该方法中，可以配置安全校验规则、设置校验结果处理器等。在5.7版本之后，这部分可以直接注入SecurityFilterChain Bean，而不需要重写configure方法。

如果想配置多个HttpSecurity，可以在SecurityConfig类外再嵌套一层。

```
@EnableWebSecurity  //开启安全注解
public class MultiHttpSecurityConfig {
    // 配置第一个HttpSecurity
    @Configuration
    // 指定顺序
    @Order(1)
    public static class SecurityConfig {
        ...
    }
    // 配置第二个HttpSecurity
    @Configuration
    public static class SecurityConfig2{
        ...
    }
}
```

以上配置完成后，基本框架就搭建完成了。

4.2 项目前期准备

4.2.1 项目需求说明

本章涉及的购物车管理系统主要包含的功能有账号管理、购物车管理、商品购买记录维护等。能够实现账号的登录、注销和注册,商品加入购物车、从购物车中移除、登记购买记录,等等。虽然此项目非常简单,但所用的框架和技术都很流行,掌握之后可以对整个项目做扩展。

4.2.2 系统功能设计

根据购物车管理系统要实现的主要功能,设计对应的功能列表。该系统的功能结构图如图4.20所示。

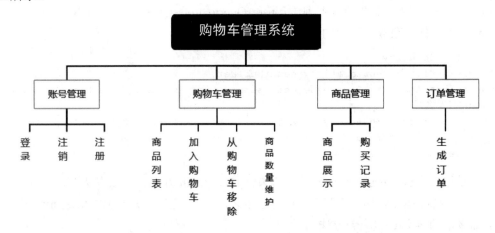

图4.20 购物车管理系统功能图

- 账号管理:实现账号的登录、注销和注册。
- 购物车管理:实现商品加入购物车、从购物车中移除商品,同步更新商品库存信息等功能。
- 商品管理:商品展示页面,维护商品购买记录。
- 订单管理:生成订单。

4.2.3 系统数据库设计

购物车管理系统主要涉及6个表,用户表(t_user)、角色表(t_role)、用户角色关联表(user_role)、商品表(t_product)、购买记录表(t_sold)、订单表(t_order)。下面我们逐一介绍这6个表。

用户表(t_user)主要用于用户登录、注销、注册,包含字段用户id(user_id)、是否激活(active)、邮箱(email)、名字(name)、姓氏(last_name)。

角色表(t_role)主要用于用户认证鉴权,包含的字段有角色id(role_id)、角色名称(role)。

用户角色关联表（user_role）主要用于用户鉴权，包含的字段有用户id（user_id）、角色id（role_id）。

商品表（t_product）主要用于商品展示和购买，包含的字段有商品id（product_id）、商品描述（description）、商品名称（name）、商品价格（price）、商品数量（quantity）。

订单表主要用于保存订单信息，包含的字段有订单id（order_id）、创建时间（create_time）、订单金额（payment）、购买用户id（user_id）。

购买记录表主要用于记录购买信息，包含的字段有记录id（sold_id）、商品数量（quantity）、对应的订单id（order_id）、对应的商品id（product_id）。

本章项目没有建表语句，全程使用Spring Data JPA托管，启动时自动生成表结构，并执行对应的sql文件初始化数据库。

4.2.4 系统文件说明

本项目系统文件的组织结构如图4.21所示。

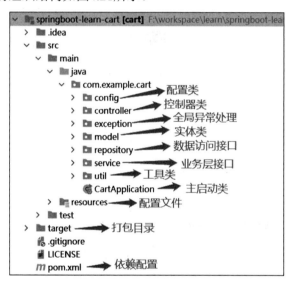

图4.21　项目后端文件组织结构

下面对各个部分做简要说明：

（1）config：Java配置类，配置一些不在配置文件中的信息，例如认证结果处理器、安全认证配置等。

（2）controller：控制器类，所有controller层的类。

（3）exception：全局异常处理。

（4）model：与数据库对应的实体类。

（5）service：业务逻辑的主要实现层，在该层可以实现事务。

（6）resources：资源或配置文件，包括Spring Boot的配置文件，这里使用的是application.properties形式；还包括一些其他文件，例如数据库初始化SQL、Thymeleaf页面模板文件夹templates、公用页面样式CSS。

（7）pom.xml：依赖JAR包，使用Maven进行项目构建，依赖JAR包统一在pom文件中进行管理。

（8）target：打包目录，构建完成后存放项目war包或JAR包的目录。

4.3 项目前端设计

4.3.1 登录

用户输入正确的用户名和密码即可完成登录。登录页面如图4.22所示。

图4.22 登录页面

以下是登录页面对应的页面代码。

```
<!DOCTYPE html>
<html xmlns="http://www.w3.org/1999/xhtml" xmlns:th="http://www.thymeleaf.org">
<head>
    <div th:replace="/fragments/header :: header"/>
</head>
<body>
<div th:replace="/fragments/header :: navbar"/>
<div class="container">
    <div class="row" style="margin-top:20px">
        <div class="col-xs-12 col-sm-8 col-md-6 col-sm-offset-2 col-md-offset-3">
            <form th:action="@{/login}" method="post">
                <fieldset>
                    <div th:if="${param.error}">
                        <div class="alert alert-danger">
                            无效的用户名或密码
                        </div>
                    </div>
                    <div th:if="${param.logout}">
                        <div class="alert alert-info">
                            已注销
                        </div>
                    </div>
                    <div class="form-group">
                        <input type="text" name="username" id="username" class="form-control input-lg" placeholder="UserName" required="true" autofocus="true"/>
```

```html
                </div>
                <div class="form-group">
                    <input type="password" name="password" id="password" class="form-control input-lg" placeholder="Password" required="true"/>
                </div>
                <div class="row">
                    <div class="col-sm-3" style="float: none; margin: 0 auto;">
                        <input type="submit" class="btn btn-primary btn-block" value="Login"/>
                    </div>
                </div>
            </fieldset>
        </form>
    </div>
</div>
<div th:replace="/fragments/footer :: footer"/>
</body>
</html>
```

整个页面使用HTML5，采用Thymeleaf模板做动态数据处理。使用xmlns引入Thymeleaf命名空间th，后续使用时采用th作为前缀。

在头部文件中，使用th:replace命令引入公共的头部信息。双冒号代表使用里边的某个组件，例如"/fragments/header :: header"代表使用fragments文件夹下的header文档中的header组件。

```html
<div th:fragment="header">
    <title th:attr="data-custom=#{thymeleaf.app.title}">Shop</title>

    <link href="http://cdn.jsdelivr.net/webjars/bootstrap/3.3.4/css/bootstrap.min.css"
          th:href="@{/webjars/bootstrap/3.3.7/css/bootstrap.min.css}"
          rel="stylesheet" media="screen"/>

    <script src="http://cdn.jsdelivr.net/webjars/jquery/2.1.4/jquery.min.js"
            th:src="@{/webjars/jquery/2.1.4/jquery.min.js}"></script>

    <link rel="stylesheet" th:href="@{/css/main.css}"
          href="../../css/main.css"/>
</div>
```

同理，导航条navBar也是同样的引用方式，即"/fragments/header :: navbar"。它在header文档中对应的组件为navBar。声明代码如下：

```html
<div th:fragment="navbar">
    <nav class="navbar navbar-inverse">
        <div class="container">
            <div class="navbar-header">
                <a class="navbar-brand" th:href="@{/home}">Shop</a>
            </div>
            <div id="navbar" class="collapse navbar-collapse navbar-right">
                <!-- 仅在用户尚未通过身份验证时显示shoppingCart -->
                <ul class="nav navbar-nav" sec:authorize="isAuthenticated()">
```

```
                    <li class="active"><a th:href="@{/shoppingCart}">购物车
</a></li>
                </ul>
                <!-- 仅在用户尚未通过身份验证时显示注册 -->
                <ul class="nav navbar-nav" sec:authorize="!isAuthenticated()">
                    <li class="active"><a th:href="@{/registration}">注册</a></li>
                </ul>
                <!-- 仅在用户尚未通过身份验证时显示登录信息 -->
                <ul class="nav navbar-nav" sec:authorize="!isAuthenticated()">
                    <li class="active"><a th:href="@{/login}">登录</a></li>
                </ul>
                <!-- 仅在用户通过身份验证时显示注销 -->
                <ul class="nav navbar-nav" sec:authorize="isAuthenticated()">
                    <li class="active"><a th:href="@{/logout}">注销</a></li>
                </ul>
            </div>
        </div>
    </nav>
</div>
```

除fragment组件外，登录页面还用到th:if条件判断语句来控制信息的显示。当用户输入的用户名或密码错误时，可以在此提示错误信息。

4.3.2 注册

用户输入正确的基本信息后，即可完成注册。注册页面如图4.23所示。

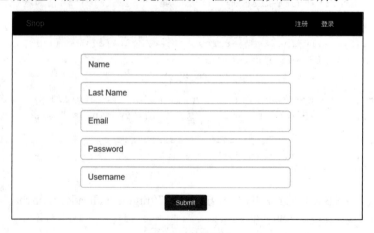

图4.23 注册页面

这里用到了Thymeleaf中的*变量。使用object命令指定user对象，然后使用*引用变量user中的属性，例如th:field="*{name}"。

```
<form autocomplete="off" action="#" th:action="@{/registration}"
      th:object="${user}" method="post" role="form">
    <div class="form-group">
        <!--th:errors结合spring validation使用。如果返回错误的字段为name，则显示该标签。下同。-->
```

```
        <label th:if= "${#fields.hasErrors('name')}" th:errors="*{name}"
            class="alert alert-danger"> </label>
        <input type="text" th:field="*{name}" placeholder="Name"
            class="form-control input-lg"/>
    </div>
    ...
</form>
```

4.3.3 商品展示页面

商品展示页面主要用来展示商品列表，包括商品名称、详细描述、价格和数量。如果当前用户已登录，则显示购买按钮，否则不显示。页面效果如图4.24所示。

图4.24　商品展示页面

这里有一个比较关键的地方，即权限控制。由当前登录状态决定是否显示购买按钮，使用的是Spring Security标签。这里的sec标签对应的是pom文件中的thymeleaf-extras-springsecurity5依赖。

```
<html xmlns="http://www.w3.org/1999/xhtml"
    xmlns:th="http://www.thymeleaf.org"
    xmlns:sec="http://www.thymeleaf.org/thymeleaf-extras-springsecurity5">
    <head>
    </head>
```

```html
        <body>
            <div th:fragment="products">
                <div class="panel-default well" th:each="product : ${products}">
                    <div class="panel-heading">
                        <h1 th:text="${product.name}"></h1>
                    </div>
                    <h3 th:text="${product.description}" class="panel-body">Description</h3>
                    <div class="row panel-footer">
                        <div th:inline="text" class="col-md-2">价格：￥ [[${product.price}]]</div>
                        <div th:inline="text" class="col-md-9">库存：[[${product.quantity}]]</div>
                        <a th:href="@{'/shoppingCart/addProduct/{id}'(id=${product.id})}" class="col-md-1"
                           sec:authorize="isAuthenticated()" th:if="${product.quantity}>0">
                            <button type="button" class="btn btn-primary" th:text="购买">购买</button>
                        </a>
                    </div>
                    <br/>
                </div>
            </div>
        </body>
    </html>
```

由xmlns:sec="http://www.thymeleaf.org/thymeleaf-extras-springsecurity5"这一句引入认证标签，前缀为sec。在按钮的a链接上加入sec:authorize="isAuthenticated()"判断是否有权限，如果没有则返回false，那么当前链接及链接内的内容不加载。

这个页面还用到了Thymeleaf的遍历标签th:each，迭代的对象是Map。th:each不仅可以遍历当前对象，还支持状态变量。基本语法是th:each="obj,iterStat : ${list}"，在这里，iterStat就是状态变量的名称，也可以自定义名称，它包含如下属性：

- index：当前遍历对象的下标索引，从0开始。
- count：当前遍历对象的计数值，从1开始。
- size：当前列表的元素总数。
- current：当前的遍历对象。
- even/odd：当前遍历对象排序是奇数还是偶数，对应布尔值为true。
- first：如果是第一个对象，该值为true。
- last：如果是最后一个对象，该值为true。

例如，根据当前状态变量奇偶排序加载不同的背景色：

```html
<tr th:each="prod,prodStat:${products}" th:class="${prodStat.odd}?'odd': 'even'">
    ...
</tr>
```

单击"购买"按钮，发送加入购物车请求，将对应商品加入购物车。这里用到了a链接，使用th:href标签动态设置按钮的值,动态取得当前商品的id作为链接参数一起传递到后台请求。

这里还用到了分页组件，可以参考后续4.3.6节公共页面中分页部分的讲解。

常用的Spring Security标签有以下4种：

- sec:authorize="isAuthenticated()"：判断用户是否已经登录认证，引号内的参数必须是isAuthenticated()。
- sec:authentication="name"：获得当前用户的用户名，引号内的参数必须是name。
- sec:authorize="hasRole('role')"：判断当前用户是否拥有指定的权限，引号内的参数为权限的名称。
- sec:authentication="principal.authorities"：获得当前用户的全部角色，引号内的参数必须是principal.authorities。

4.3.4 购物车页面

在商品页面单击"购买"按钮或者直接单击导航条中的"购物车"按钮，跳转至购物车页面，页面如图4.25所示。

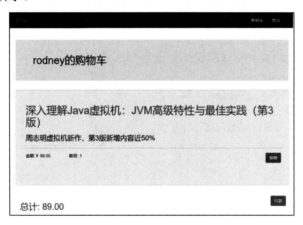

图4.25 购物车页面

当前页面同样有Spring Security的权限控制，用到的同样是上文中提到的sec标签。获取当前登录用户的用户名，如果当前用户未登录或认证已失效，那么跳转至登录页。

```
<h1 class="jumbotron">
    <span sec:authentication="name"></span>的购物车
</h1>
```

当单击"移除"按钮时，触发a标签绑定的链接，并将当前商品的id传递过去，完成移除请求。

```
<a th:href="@{'/shoppingCart/removeProduct/{id}'(id=${product.getKey().id})}" class="col-md-1">
    <button type="button" class="btn btn-primary" th:text="移除">Remove</button>
</a>
```

同理，单击"付款"按钮时，根据a标签绑定的链接发送付款请求。这里模拟付款，当单击"付款"按钮后，将当前购物车内的商品添加到订单，并清空购物车。

4.3.5 通用导航

通用导航是整个网站公用的头部信息。公共头部信息包括公共header（例如标题title）、引入的JS和CSS、导航条navbar、菜单等信息。在header文件中，使用fragment命令声明header和navbar组件，便于灵活引入其他页面。在footer文件中，同样使用fragment命令声明footer组件。通常一个导航应该包含网站的Logo、名称、导航菜单、当前登录用户信息、常用按钮以及消息提示等，但我们演示的这个购物车管理系统比较简单，所以这里只做了粗略划分，显示了网站名称和部分按钮。如果用户未登录，则应该显示登录、注册按钮，如图4.26所示。

图4.26 未登录导航

如果用户已登录，则应该显示注销、购物车按钮，如图4.27所示。

图4.27 已登录导航

以上页面通过Spring Security权限控制标签即可实现，代码如下：

```html
<div th:fragment="navbar">
    <nav class="navbar navbar-inverse">
        <div class="container">
            <div class="navbar-header">
                <a class="navbar-brand" th:href="@{/home}">Shop</a>
            </div>
            <div id="navbar" class="collapse navbar-collapse navbar-right">
                <!-- 仅在用户尚未通过身份验证时显示shoppingCart -->
                <ul class="nav navbar-nav" sec:authorize="isAuthenticated()">
                    <li class="active"><a th:href="@{/shoppingCart}">购物车</a></li>
                </ul>
                <!-- 仅在用户尚未通过身份验证时显示注册 -->
                <ul class="nav navbar-nav" sec:authorize="!isAuthenticated()">
                    <li class="active"><a th:href="@{/registration}">注册</a></li>
                </ul>
                <!-- 仅在用户尚未通过身份验证时显示登录信息 -->
                <ul class="nav navbar-nav" sec:authorize="!isAuthenticated()">
                    <li class="active"><a th:href="@{/login}">登录</a></li>
                </ul>
                <!-- 仅在用户通过身份验证时显示注销 -->
                <ul class="nav navbar-nav" sec:authorize="isAuthenticated()">
                    <li class="active"><a th:href="@{/logout}">注销</a></li>
                </ul>
            </div>
        </div>
    </nav>
</div>
```

4.3.6 通用分页

整个项目页面中用到分页的地方全部采用统一分页。这里的参数需要和后端定义的分页类对应。使用fragment标签声明分页组件，命名为pagination，供需要分页的页面引用。

```html
<div th:fragment="pagination">
    <!-- 翻页 -->
    <div class="pagination" th:with="baseUrl=${URLparameter}">
        <span th:if="${pager.indexOutOfBounds()}">
            页面超出范围，回到
            <a class="pageLink" th:href="@{${baseUrl}(page=1)}">Home</a>.
        </span>
        <span th:unless="${pager.indexOutOfBounds()}">
            <span th:if="${pager.hasPrevious()}">
                <a class="pageLink" th:href="@{${baseUrl}(page=1)}"> &laquo; first</a>
                <a class="pageLink" th:href="@{${baseUrl}(page=${pager.getPageIndex() - 1})}"> previous</a>
            </span>
            <span th:if="${pager.getTotalPages() != 1}" th:text= "'Page ' + ${pager.getPageIndex()} + ' of ' + ${pager.getTotalPages()} + '.'">
            </span>
            <span th:if="${pager.hasNext()}">
                <a class="pageLink" th:href="@{${baseUrl}(page=${pager.getPageIndex() + 1})}">next</a>
                <a class="pageLink" th:href="@{${baseUrl}(page=${pager.getTotalPages()})}">last &raquo;</a>
            </span>
        </span>
    </div>
</div>
```

根据后台返回的pager对象来判断当前的分页情况。这里使用th:with定义局部变量baseUrl，这个值代表分页出现错误调用时跳转的页面，例如分页出现错误时，跳转至home页，在引入该分页组件时，可以传递URLparameter参数：

```html
<div th:replace="/fragments/pagination :: pagination(URLparameter='/home')"/>
```

接下来，根据当前页和总页数判断按钮的显示情况。存在前一页的话，显示previous按钮，存在后一页的话，显示next按钮，并同时判断是否显示跳转第一页和最后一页的按钮。

4.3.7 安全校验错误页面

安全校验错误页面用来提示用户当前没有该页面的操作权限。

```html
<div class="container">
    <div class="starter-template">
        <h1>403 - Access is denied</h1>
        <div th:inline="text">Hello '[[${#httpServletRequest.remoteUser}]]',
```

```
            没有访问该页面的权限
        </div>
    </div>
</div>
```

4.4 项目后端实现

4.4.1 登录与登录认证

1. 登录校验

登录校验，我们这里使用Spring Security托管，无须再单独编写登录逻辑，只需做好以下配置即可。

首先，在配置文件中配置数据库查询语句：

```
spring.queries.users-query=select username, password, active from t_user where username=?
```

然后在SecurityConfig配置类中设置登录页的URL：

```
//登录页
.loginPage("/login")
//默认成功返回URL
.defaultSuccessUrl("/home")
```

最后在登录相关的Controller中简单写明跳转链接：

```
@GetMapping("/login")
public String login(Principal principal) {
    if (principal != null) {              //存在认证信息
        return "redirect:/home";          //直接跳转首页
    }
    return "/login";                      //返回登录页面
}
```

Principal对象是我们在4.1.1节中讲到的认证对象Authentication中的用户信息，如果该对象存在，则表明当前用户已通过认证，直接跳转到首页即可。如果不存在该对象，那么认证失败，需要重新登录。下面重点看Spring Security的登录认证过程。

2. 登录认证

登录鉴权涉及的过滤器主要是UsernamePasswordAuthenticationFilter。前面提到，默认情况下Spring Security会提供一个ProviderManager注入过滤器对应的AuthenticationManager属性中，这个ProviderManager主要就是实现认证工作的。而AuthenticationManager的初始化分为两种，一种是全局的AuthenticationManager，另一种是局部的AuthenticationManager。

Spring Security允许存在多条过滤器链，每一条过滤器链隶属于一个HttpSecurity，所以我们可以同时配置多个HttpSecurity。而每一个HttpSecurity又对应一个AuthenticationManager，这个就是局部的AuthenticationManager，它们都有一个共同的parentAuthenticationManager属性，就是全局的AuthenticationManager。

从前面的框架搭建部分可以看到，本章项目中我们在SecurityConfig文件中配置了全局的AuthenticationManagerBuilder对象，定义了用户查询语句、角色查询语句和密码加密方式，那么在注入ProviderManager对象时就会自动设置这些属性。

AuthenticationManagerBuilder顾名思义就是用来构建AuthenticationManager的。下面我们来看一下它当中的部分关键代码。

```java
public class AuthenticationManagerBuilder extends
AbstractConfiguredSecurityBuilder<AuthenticationManager,
AuthenticationManagerBuilder>
        implements ProviderManagerBuilder<AuthenticationManagerBuilder> {
    ...
    public AuthenticationManagerBuilder parentAuthenticationManager(
            AuthenticationManager authenticationManager) {
        if (authenticationManager instanceof ProviderManager) {
            eraseCredentials(((ProviderManager) authenticationManager)
                    .isEraseCredentialsAfterAuthentication());
        }
        this.parentAuthenticationManager = authenticationManager;
        return this;
    }
    public InMemoryUserDetailsManagerConfigurer<AuthenticationManagerBuilder>
inMemoryAuthentication() throws Exception {
        return apply(new InMemoryUserDetailsManagerConfigurer<>());
    }
    public JdbcUserDetailsManagerConfigurer<AuthenticationManagerBuilder>
jdbcAuthentication() throws Exception {
        return apply(new JdbcUserDetailsManagerConfigurer<>());
    }
    public <T extends UserDetailsService> DaoAuthenticationConfigurer
<AuthenticationManagerBuilder, T> userDetailsService(
            T userDetailsService) throws Exception {
        this.defaultUserDetailsService = userDetailsService;
        return apply(new DaoAuthenticationConfigurer<>(
                userDetailsService));
    }
    @Override
    protected ProviderManager performBuild() throws Exception {
        if (!isConfigured()) {
            logger.debug("No authenticationProviders and no
parentAuthenticationManager defined. Returning null.");
            return null;
        }
        ProviderManager providerManager = new
ProviderManager(authenticationProviders,
                parentAuthenticationManager);
        if (eraseCredentials != null) {
```

```
                providerManager.setEraseCredentialsAfterAuthentication
(eraseCredentials);
            }
            if (eventPublisher != null) {
                providerManager.setAuthenticationEventPublisher(eventPublisher);
            }
            providerManager = postProcess(providerManager);
            return providerManager;
        }
    }
```

通过调用parentAuthenticationManager方法给当前这个构造器要构造的AuthenticationManager对象设置一个parentAuthenticationManager。inMemoryAuthentication、jdbcAuthentication以及userDetailsService方法的作用是配置数据源。performBuild方法的作用就是根据当前AuthenticationManagerBuilder来构建一个AuthenticationManager。AuthenticationManager本身是一个接口，它的默认实现是ProviderManager，所以这里构建的就是ProviderManager。在构建ProviderManager时，第一个参数传入authenticationProviders，即该ProviderManager所管理的所有的AuthenticationProvider，第二个参数传入ProviderManager的parentAuthenticationManager。

在创建AuthenticationManager实例化对象前，还要了解一个配置类，叫AuthenticationConfiguration，它主要提供一些全局的初始化Bean对象，例如AuthenticationManagerBuilder、UserDetailService、AuthenticationProvider等。AuthenticationConfiguration的源码如下：

```
    @Configuration(proxyBeanMethods = false)
    @Import(ObjectPostProcessorConfiguration.class)
    public class AuthenticationConfiguration {
        @Bean
        // 构造器
        public AuthenticationManagerBuilder authenticationManagerBuilder(
                ObjectPostProcessor<Object> objectPostProcessor, ApplicationContext
context) {
            //延迟密码编码
            LazyPasswordEncoder defaultPasswordEncoder = new
LazyPasswordEncoder(context);
            //认证事件发布者
            AuthenticationEventPublisher authenticationEventPublisher =
getBeanOrNull(context, AuthenticationEventPublisher.class);
            DefaultPasswordEncoderAuthenticationManagerBuilder result = new
DefaultPasswordEncoderAuthenticationManagerBuilder(objectPostProcessor,
defaultPasswordEncoder);
            if (authenticationEventPublisher != null) {
                result.authenticationEventPublisher(authenticationEventPublisher);
            }
            return result;
        }
        // 全局认证配置适配器
        @Bean
```

```java
        public static GlobalAuthenticationConfigurerAdapter
enableGlobalAuthenticationAutowiredConfigurer(
                ApplicationContext context) {
            return new EnableGlobalAuthenticationAutowiredConfigurer(context);
        }
        // 初始化用户信息管理器配置类
        @Bean
        public static InitializeUserDetailsBeanManagerConfigurer
initializeUserDetailsBeanManagerConfigurer(ApplicationContext context) {
            return new InitializeUserDetailsBeanManagerConfigurer(context);
        }
        // 初始化认证提供者管理器配置类
        @Bean
        public static InitializeAuthenticationProviderBeanManagerConfigurer
initializeAuthenticationProviderBeanManagerConfigurer(ApplicationContext context) {
            return new InitializeAuthenticationProviderBeanManagerConfigurer(context);
        }
        // 初始化AuthenticationManager
        public AuthenticationManager getAuthenticationManager() throws Exception {
            if (this.authenticationManagerInitialized) {
                return this.authenticationManager;
            }
            AuthenticationManagerBuilder authBuilder =
this.applicationContext.getBean(AuthenticationManagerBuilder.class);
            if (this.buildingAuthenticationManager.getAndSet(true)) {
                return new AuthenticationManagerDelegator(authBuilder);
            }
            for (GlobalAuthenticationConfigurerAdapter config : globalAuthConfigurers) {
                authBuilder.apply(config);
            }
            authenticationManager = authBuilder.build();
            if (authenticationManager == null) {
                authenticationManager = getAuthenticationManagerBean();
            }
            this.authenticationManagerInitialized = true;
            return authenticationManager;
        }
        @Autowired
        public void setApplicationContext(ApplicationContext applicationContext) {
            this.applicationContext = applicationContext;
        }
        @Autowired
        public void setObjectPostProcessor(ObjectPostProcessor<Object>
objectPostProcessor) {
            this.objectPostProcessor = objectPostProcessor;
        }
        // 启用全局配置自动配置的适配器
        private static class EnableGlobalAuthenticationAutowiredConfigurer extends
                GlobalAuthenticationConfigurerAdapter {
            private final ApplicationContext context;
            private static final Log logger = LogFactory
```

```
            .getLog(EnableGlobalAuthenticationAutowiredConfigurer.class);
    EnableGlobalAuthenticationAutowiredConfigurer(ApplicationContext context) {
        this.context = context;
    }
    @Override
    public void init(AuthenticationManagerBuilder auth) {
        Map<String, Object> beansWithAnnotation = context
            .getBeansWithAnnotation(EnableGlobalAuthentication.class);
        if (logger.isDebugEnabled()) {
            logger.debug("Eagerly initializing " + beansWithAnnotation);
        }
    }
}
```

首先构建了一个AuthenticationManagerBuilder实例，这个实例就是用来构建全局AuthenticationManager的AuthenticationManagerBuilder，具体的构建过程在getAuthenticationManager方法中。

initialize方法用来构建全局的UserDetailService和AuthenticationProvider。正常情况下是不会用到这几个Bean的，只有当getAuthenticationManager方法被调用时，这些默认的Bean才会被配置，而getAuthenticationManager方法被调用，意味着我们要使用系统默认配置的AuthenticationManager作为parentAuthenticationManager，而在实际使用中，我们一般不会使用系统默认配置的AuthenticationManager作为parentAuthenticationManager，大部分情况都会自定义一些内容。

熟悉了源码之后，我们来看下构造器何时创建AuthenticationManager的实例化对象。在5.7版本之前，这里要用到一个适配器WebSecurityConfigurerAdapter。在这个适配器中有三个重要的方法涉及AuthenticationManager的初始化问题，第一个方法是setApplicationContext。

```
public void setApplicationContext(ApplicationContext context) {
    this.context = context;
    ObjectPostProcessor<Object> objectPostProcessor = context.getBean
(ObjectPostProcessor.class);
    LazyPasswordEncoder passwordEncoder = new LazyPasswordEncoder(context);
    authenticationBuilder = new
DefaultPasswordEncoderAuthenticationManagerBuilder(objectPostProcessor,
passwordEncoder);
    localConfigureAuthenticationBldr = new
DefaultPasswordEncoderAuthenticationManagerBuilder(objectPostProcessor,
passwordEncoder) {
        @Override
        public AuthenticationManagerBuilder eraseCredentials(boolean
eraseCredentials) {
            authenticationBuilder.eraseCredentials(eraseCredentials);
            return super.eraseCredentials(eraseCredentials);
        }
        @Override
        public AuthenticationManagerBuilder authenticationEventPublisher
(AuthenticationEventPublisher eventPublisher) {
```

```
            authenticationBuilder.authenticationEventPublisher(eventPublisher);
            return super.authenticationEventPublisher(eventPublisher);
        }
};}
```

首先创建 AuthenticationManagerBuilder 的实例化对象，这里和前面说到的 AuthenticationConfiguration 中创建的 AuthenticationManagerBuilder 是相辅的，如果由 WebSecurityConfigurerAdapter 创建的对象被禁用的话，会默认使用前面 AuthenticationConfiguration中创建的AuthenticationManagerBuilder实例化对象。在5.7版本之后，WebSecurityConfigurerAdapter 已被标注为过时。我们可以直接在SecurityConfig中注入所需要的AuthenticationManager对象。可以参考前面4.1.4节框架搭建部分的讲解。

第二个方法是getHttp。

```
protected final HttpSecurity getHttp() throws Exception {
    if (http != null) {
        return http;
    }
    AuthenticationEventPublisher eventPublisher = getAuthenticationEventPublisher();
    localConfigureAuthenticationBldr.authenticationEventPublisher(eventPublisher);
    AuthenticationManager authenticationManager = authenticationManager();
    authenticationBuilder.parentAuthenticationManager(authenticationManager);
    Map<Class<?>, Object> sharedObjects = createSharedObjects();
    http = new HttpSecurity(objectPostProcessor, authenticationBuilder,
            sharedObjects);
    ...
    return http;}
```

这个方法主要实现为HttpSecurity 对象设置parentAuthenticationManager属性。

第三个方法是authenticationManager。

```
protected AuthenticationManager authenticationManager() throws Exception {
    if (!authenticationManagerInitialized) {
        configure(localConfigureAuthenticationBldr);
        if (disableLocalConfigureAuthenticationBldr) {
            authenticationManager = authenticationConfiguration
                .getAuthenticationManager();
        }
        else {
            authenticationManager = localConfigureAuthenticationBldr.build();
        }
        authenticationManagerInitialized = true;
    }
    return authenticationManager;}protected void
configure(AuthenticationManagerBuilder auth) throws Exception {
    this.disableLocalConfigureAuthenticationBldr = true;
}
```

这个方法主要是初始化一个AuthenticationManager对象，并标记已初始化该对象。

在5.7版本之前，一般情况下会在自定义的SecurityConfig 类中重写这个方法，也就是前面所说的会自定义AuthenticationManager，也可以直接使用默认的，这时就会调用if代码块中的内容。

如果重写了configure方法，那默认的AuthenticationManagerBuilder就被更改了，会调用configure(localConfigureAuthenticationBldr)加载自定义配置的内容，然后使用build方法构建新的AuthenticationManager对象，否则的话就调用authenticationConfiguration.getAuthenticationManager()获取默认的AuthenticationManager。在5.7之后的新版本中，我们可以直接注入。虽然版本不同，使用方法不同，但是大体流程和原理基本相同。

所以使用UsernamePasswordAuthenticationFilter过滤器做登录认证的流程大致如图4.28所示。

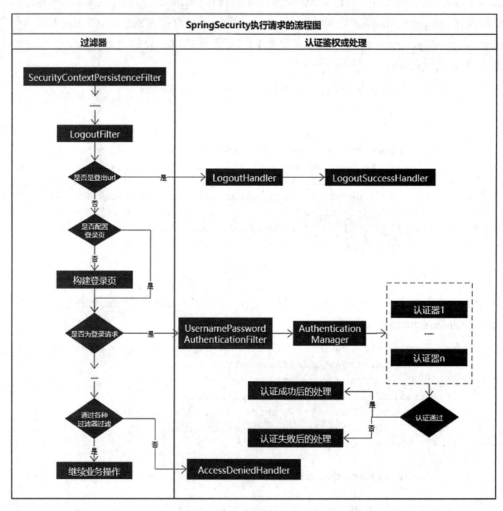

图4.28　Spring Security工作流程图

流程描述：客户端发起一个请求，进入过滤器链。当到达LogoutFilter的时候判断是否是注销路径，如果是注销路径则到LogoutHandler，如果注销成功则到LogoutSuccessHandler进行注销成功处理；如果不是注销路径则直接进入下一个过滤器。当到达UsernamePasswordAuthenticationFilter的时候判断是否为登录路径，如果是，则进入该过滤器进行登录操作，如果登录失败则到AuthenticationFailureHandler登录失败处理器进行处理，如果登录成功则到AuthenticationSuccessHandler登录成功处理器进行处理；如果不是登录请求则进入后续过滤器。

当到FilterSecurityInterceptor的时候会拿到URI，根据URI去找对应的鉴权管理器做鉴权工作，鉴权成功则到Controller层继续处理业务，否则到AccessDeniedHandler鉴权失败处理器进行处理。

4.4.2 注册与参数验证

注册时，需要传递用户的基本信息，包含用户名、邮箱、密码、名字、姓氏等。后台获取到前端传递的参数后，首先校验参数的合法性，先使用@Valid注解，再使用Service方法校验邮箱和用户名是否已存在，如果存在则给出已存在提示，注册失败，重新返回注册页；如果不存在，则向数据库中新增一条用户信息。

```
@PostMapping(value = "/registration")
public ModelAndView createNewUser(@Valid User user, BindingResult bindingResult) {
    if (userService.findByEmail(user.getEmail()).isPresent()) {
        bindingResult.rejectValue("email", "error.user",
                "已经有一个用户注册了所提供的电子邮件");
    }
    if (userService.findByUsername(user.getUsername()).isPresent()) {
        bindingResult.rejectValue("username", "error.user",
                "已经有一个使用提供的用户名注册的用户");
    }
    ModelAndView modelAndView = new ModelAndView();
    if (bindingResult.hasErrors()) {
        //注册失败，返回错误信息
        modelAndView.setViewName("/registration");
    } else {
        // 注册成功
        userService.saveUser(user);
        modelAndView.addObject("successMessage", "用户已成功注册");
        modelAndView.addObject("user", new User());
        modelAndView.setViewName("/registration");
    }
    return modelAndView;
}
```

校验参数合法性时，使用的是@Valid注解和BindingResult类。@Valid是Java Validation提供的标准验证注解，Spring对此提供了支持。当添加了@Valid注解后，对应的参数实体类就可以通过标准的验证注解设置验证规则，例如User类中email邮箱字段不能为空、格式必须为Email等。设置如下：

```
public class User implements Serializable {
    ...
    @Column(name = "email", unique = true, nullable = false)
    @Email(message = "*请提供有效的电子邮件")
    @NotBlank(message = "*请提供电子邮件")
    private String email;
    @Column(name = "password", nullable = false)
    @Length(min = 5, message = "*您的密码必须至少包含5个字符")
    @NotEmpty(message = "*请提供您的密码")
    @JsonIgnore
    private String password;
```

```
    @Column(name = "username", nullable = false, unique = true)
    @Length(min = 5, message = "*您的用户名必须至少包含5个字符")
    @NotEmpty(message = "*请提供您的名字")
    private String username;
    @Column(name = "name")
    @NotEmpty(message = "*请提供您的名字")
    private String name;
    @Column(name = "last_name")
    @NotEmpty(message = "*请提供您的姓氏")
    private String lastName;
    ...
}
```

Java提供的关于验证的标准注解遵循Bean Validation 规范。这个标准随着JSR（Java Specification Requests）标准提出，代表Bean Validation。Bean Validation 是一个运行时的数据验证框架，在验证之后验证的错误信息会被马上返回。但是在Java中这仅仅是一个规范，并没有具体的实现。在这一方面Hibernate-Validator框架提供了规范中所有内置验证的实现，还额外做了扩展。

Bean Validation规范目前有3个版本，表4.3是其版本信息和对应的Hibernate-Validator框架的版本信息。

表4.3　Bean Validation规范

版　　本	发行年份	对应的JSR版本	Hibernate实现版本	apache实现版本
Bean Validation 1.0	2009年	JSR 303	Hibernate-Validator-4.3.1.Final	org.apache.bval 0.5
Bean Validation 1.1	2013年	JSR 349	Hibernate-Validator-5.1.1.Final	org.apache.bval 1.1.2
Bean Validation 2.0	2017年	JSR 380	Hibernate-Validator-6.0.1.Final	org.apache.bval 2.0.3

Spring也有对应的验证框架，Spring Boot把该框架做了封装，使用时需要在pom文件中引入对应的依赖：

```
<dependency>
    <groupId>org.springframework.boot</groupId>
    <artifactId>spring-boot-starter-validation</artifactId>
</dependency>
```

最新的Bean Valadition规范支持的标准注解如表4.4所示。

表4.4　Bean Validation注解

注解名称	作　　用
@Null	限制只能为null
@NotNull	限制必须不为null
@AssertFalse	限制必须为false
@AssertTrue	限制必须为true
@DecimalMax(value)	限制必须为一个不大于指定值的数字
@DecimalMin(value)	限制必须为一个不小于指定值的数字

(续表)

注解名称	作　用
@Digits(integer,fraction)	限制必须为一个小数，且整数部分的位数不能超过integer，小数部分的位数不能超过fraction
@Future	限制必须是一个将来的日期
@Max(value)	限制必须为一个不大于指定值的数字
@Min(value)	限制必须为一个不小于指定值的数字
@Past	限制必须是一个过去的日期
@Pattern(value)	限制必须符合指定的正则表达式
@Size(max,min)	限制字符长度必须在min到max之间
@Null	限制只能为null
@NotNull	限制必须不为null
@AssertFalse	限制必须为false
@AssertTrue	限制必须为true
@DecimalMax(value)	限制必须为一个不大于指定值的数字
@DecimalMin(value)	限制必须为一个不小于指定值的数字
@Digits(integer,fraction)	限制必须为一个小数，且整数部分的位数不能超过integer，小数部分的位数不能超过fraction
@Future	限制必须是一个将来的日期
@Max(value)	限制必须为一个不大于指定值的数字
@Min(value)	限制必须为一个不小于指定值的数字
@Past	限制必须是一个过去的日期
@Pattern(value)	限制必须符合指定的正则表达式
@Size(max,min)	限制字符长度必须在min到max之间
@Past	限制日期，过去的日期
@Future	限制日期，将来的日期
@Email	限制字符串，邮箱
@NotEmpty	限制不能为空，一般用于集合
@NotBlank	限制不能为空，一般用于字符串
@Positive	限制数字，正数
@PositiveOrZero	限制数字，正数或0
@Negative	限制数字，负数
@NegativeOrZero	限制数字，负数或0
@PastOrPresent	限制过去或现在的日期
@FutureOrPresent	限制将来或现在的日期

上面的代码中只讲了@NotBlank验证邮箱不能为空，关于其他注解的使用可以参考本书配套源码，也可参考Hibernate-Validator或者Spring Validation官方文档。

向数据库的t_user表增加一条新用户数据，采用的是SpringJPA框架，这部分在第3章中已经有了详细介绍，在此不再详述。

前面章节未提到的是，这里的实体类采用了多对多或一对多注解，例如实体类User中，用户与角色是多对多关系，而用户与订单则是一对多的关系，就可以像下面的代码一样使用@ManyToMany注解或@OneToMany注解。

```java
    @ManyToMany(cascade = CascadeType.ALL)
    @JoinTable(name = "user_role", joinColumns = @JoinColumn(name = "user_id"),
inverseJoinColumns = @JoinColumn(name = "role_id"))
    private Collection<Role> roles;
    @OneToMany(fetch = FetchType.LAZY,mappedBy = "user")
    private List<Order> orderList;
```

这样设置之后，可以一并查出用户对应的角色和用户对应的订单。cascade = CascadeType.ALL代表的是级联操作，即对当前用户进行增加、删除和更新操作时，是否将同样的操作应用到用户与角色对应关系上。这里设置的是所有操作，即查询、修改、删除都将一并操作。CascadeType支持的级联方式有如下6种：

- CascadeType.REMOVE：级联删除操作。删除当前实体时，与它有映射关系的实体也会跟着被删除。
- CascadeType.MERGE：级联更新（合并）操作。当Student中的数据改变时，会相应地更新Course中的数据。
- CascadeType.DETACH：级联脱管/游离操作。如果要删除一个实体，但是它有外键从而无法删除，就需要这个级联权限了，它会撤销所有相关的外键关联。
- CascadeType.REFRESH：级联刷新操作。假设场景有一个订单，订单里面关联了许多商品，这个订单可以被很多人操作，如果A对此订单和关联的商品进行了修改，同时B也进行了相同的操作，但是B比A先一步保存了数据，那么当A保存数据的时候，就需要先刷新订单信息及关联的商品信息，再保存订单及商品。
- CascadeType.PERSIST：级联保存。新增当前实体时，如果有映射关系的实体在数据库中不存在的话，会对应地同步新增实体。
- CascadeType.ALL拥有以上所有级联操作权限。

4.4.3 异常处理

1. GlobalExceptionHandler

顾名思义，GlobalExceptionHandler是一个全局的异常处理。在实际项目中，我们经常遇到需要自己处理异常的情况，如果每一个异常都单独处理，代码会琐碎杂乱，维护起来也非常低效。针对这种情况，我们在业务层代码中可以统一将异常抛出，然后再全局统一捕获处理，这样既高效又简洁，代码也易于维护。GlobalExceptionHandler类就是本项目中自定义的全局异常处理类，它的代码如下：

```java
    @ControllerAdvice
    public class GlobalExceptionHandler {
        @ExceptionHandler(Throwable.class)
        @ResponseStatus(HttpStatus.INTERNAL_SERVER_ERROR)
        public ModelAndView exception(final Throwable throwable, final Model model) {
```

```
        log.error("Exception during execution of SpringSecurity application",
throwable);
        ModelAndView modelAndView = new ModelAndView("/error");
        String errorMessage = (throwable != null ? throwable.toString() : "Unknown
error");
        modelAndView.addObject("errorMessage", errorMessage);
        return modelAndView;
    }
}
```

@ControllerAdvice 定义了拦截的规则，它是Spring3.2之后提供的新注解，需要结合@ExceptionHandler、@InitBinder 或 @ModelAttribute注解使用。比较常用的是和@ExceptionHandler注解结合，即本示例中所讲解的情况。ControllerAdvice 提供了多种指定Advice规则的定义方式，默认为空，表示适用于所有的Controller。如果要指定某些Controller适用，可以在注解后标明。标明的方式有多种，可以使用包名，也可以使用注解。例如指定包名，使用basePackages属性，该属性为默认属性，只指定该属性时，属性名可省略，即 @ControllerAdvice("cn.my.pkg") 或者@ControllerAdvice(basePackages="cn.my.pkg")，表示匹配cn.my.pkg包及其子包下的所有Controller；也可以用数组的形式指定，如@ControllerAdvice(basePackages={"cn.my.pkg", "cn.my.pkg2"})。还可以通过指定注解来匹配，比如自定义一个@MyAnnotation注解，@ControllerAdvice（annotations={CustomAnnotation.class}）表示匹配所有被这个注解修饰的 Controller。

```
@Target(ElementType.TYPE)
@Retention(RetentionPolicy.RUNTIME)
@Documented
@Component
public @interface ControllerAdvice {
    @AliasFor("basePackages")
    String[] value() default {};
    @AliasFor("value")
    String[] basePackages() default {};
    Class<?>[] basePackageClasses() default {};
    Class<?>[] assignableTypes() default {};
    Class<? extends Annotation>[] annotations() default {};
}
```

与@ControllerAdvice相对的还有一个@RestControllerAdvice，与@Controller和@RestController类似，@RestControllerAdvice比@ControllerAdvice多一个@ResponseBody注解。

@ExceptionHandler注解一般用来自定义异常处理。

```
@Target(ElementType.METHOD)
@Retention(RetentionPolicy.RUNTIME)
@Documented
public @interface ExceptionHandler {
    // 指定需要捕获的异常的Class类型
    Class<? extends Throwable>[] value() default {};
}
```

从源码可以看出，它只有一个参数，也是默认参数。这个参数接收的是class数组。本章这里捕获的是Throwable类异常，即所有抛出的异常全部通过这个注解标注的方法来处理——

@ExceptionHandler(Throwable.class)。这是非常宽泛的处理方式，对用户来说非常不友好，但限于篇幅我们这里只讲原理和方法，读者可以自己进行扩展。我们将异常划分得越详细，这里捕获的异常就会越精准。例如，数字转换异常的捕获处理：

```
@ExceptionHandler(NumberFormatException.class)      //申明捕获哪个异常类
public String NumberFormatExceptionDemo(Exception e) {
    logger.error(e.getMessage(), e);
    return "数字转换异常返回";
}
```

除了@ExceptionHandler，@ControllerAdvice还可以和@ModelAttribute、@InitBinder注解结合使用。

- @ExceptionHandler：全局异常处理。处理对应异常。
- @ModelAttribute：全局数据绑定。将参数绑定到全局，每一个注解了@RequestMapping的方法都可以获得此参数。
- @InitBinder：全局数据预处理。将参数进行预处理，例如特殊处理字符串，或预转换时间日期格式等。

```
@ControllerAdvice
public class GlobalController{
    //（1）全局数据绑定
    @ModelAttribute
    public void addUser(Model model) {
        model.addAttribute("msg", "注解@RequestMapping的方法都可以拿到此model参数");
    }
    //（2）全局数据预处理
    @InitBinder("user")
    public void initBinder(WebDataBinder binder) {
    }
    //（3）全局异常处理
    @ExceptionHandler(Exception.class)
    public String handleException(Exception e) {
        return "error";
    }
}
```

2. 自定义异常

在项目中，可以自定义异常来统一处理提示信息，方便业务处理。例如，本项目中使用的库存不足异常NotEnoughProductsInStockException。

```
public class NotEnoughProductsInStockException extends Exception {
    private static final String DEFAULT_MESSAGE = "库存产品不足";
    public NotEnoughProductsInStockException() {
        super(DEFAULT_MESSAGE);
    }
    public NotEnoughProductsInStockException(Product product) {
```

```
            super(String.format("%s 库存不足。仅剩 %d 件。", product.getName(),
product.getQuantity())));
        }
    }
```

所有遇到需要校验库存信息的地方都可以使用该异常，直接抛出，而不需要单独去处理每一个逻辑。例如，生成订单时，在ShoppingCartService中校验库存信息，如果库存不足，则直接抛出该异常。在所有调用该业务方法的地方捕捉这个异常，返回提示信息即可。

```
public void checkout(User user) throws NotEnoughProductsInStockException {
    ...
    product = one.get();
    if (product.getQuantity() < quantity) {//库存数量少于购物车中的数量
        throw new NotEnoughProductsInStockException(product);
    }
    ...
}
```

在Controller中调用时，捕捉异常，设置提示。

```
@GetMapping("/shoppingCart/checkout")
public ModelAndView checkout(Principal principal) {
    // 获取当前登录用户
    Optional<User> optionalUser =
userRepository.findByUsername(principal.getName());
    if (!optionalUser.isPresent()) {
        throw new HttpClientErrorException(HttpStatus.UNAUTHORIZED);
    }
    try {
        // 生成订单
        shoppingCartService.checkout(optionalUser.get());
    } catch (NotEnoughProductsInStockException e) {
        // 商品库存不足
        return shoppingCart().addObject("outOfStockMessage", e.getMessage());
    }
    return shoppingCart();
}
```

同理，读者可以根据业务自行设置多个异常，分门别类地对异常进行细化。这里不再详述。

4.4.4 安全校验

前面4.1.4的框架搭建中已经讲了Spring Security的配置方法和代码示例。这里讲一下每一项配置的意义。antMatchers().permitAll()所配置的是所有不需要进行安全校验的链接，这里包括所有的静态资源，例如CSS、JS或图片等。loginPage("/login").permitAll()配置的是登录页面，配置之后登录页面也不会进入安全校验。defaultSuccessUrl("/home").permitAll()配置的是默认成功返回的URL。exceptionHandling().accessDeniedHandler(accessDeniedHandler)配置的是未通过安全认证的处理器。CustomAccessDeniedHandler用来处理未通过安全校验的请求。CustomAccessDeniedHandler类继承自AccessDeniedHandler。源代码示例如下：

```java
@Component
public class CustomAccessDeniedHandler implements AccessDeniedHandler {

    @Override
    public void handle(HttpServletRequest httpServletRequest,
                       HttpServletResponse httpServletResponse,
                       AccessDeniedException e) throws IOException {
        Authentication auth = SecurityContextHolder.getContext().getAuthentication();
        if (auth != null) {
            log.info(String.format("User '%s' attempted to access the protected URL: %s", auth.getName(), httpServletRequest.getRequestURI()));
        }
        httpServletResponse.sendRedirect(httpServletRequest.getContextPath() + "/403");
    }
}
```

由于默认的AccessDeniedHandler中,对于拦截到的请求所提示的信息可阅读性不强,所以这里可以重写handle方法,自定义提示信息以及发生错误之后需要跳转的页面。

错误页面可以使用专门的Controller来定义。CustomErrorController定义了错误页面的跳转路径。

```java
@RestController
public class CustomErrorController {
    @GetMapping("/403")
    public ModelAndView error403() {
        return new ModelAndView("/403");
    }
}
```

4.4.5 商城首页

商城首页需要的数据是商品列表。传入的参数是分页信息,从商品表中取出所有商品信息分页展示,包括商品基本信息、库存信息、价格信息等。

```java
@GetMapping(value = {"/home", "/"})
public ModelAndView home(@RequestParam(value = "page", defaultValue = "1") Integer page) {
    //实际分页的参数与页面显示的页数不同,需要在页数的基础上减1
    int evalPage = page - 1;
    //分页查询
    Page<Product> products = productRepository.findAll(PageRequest.of(evalPage, 5));
    //使用工具类从分页查询中提取分页组件所需的数据
    Pager pager = new Pager(products);
    ModelAndView modelAndView = new ModelAndView();
    modelAndView.addObject("products", products);
    modelAndView.addObject("pager", pager);
    modelAndView.setViewName("/home");
    return modelAndView;
}
```

这里的分页使用的是Spring Data JPA中的分页功能。查询商品列表findAll()用到的是Spring JPA默认提供的方法。这个方法是接口PagingAndSortingRepository中的默认方法，使用时传入Pageable分页对象即可。

```
@NoRepositoryBean
public interface PagingAndSortingRepository<T, ID> extends CrudRepository<T, ID> {
    Iterable<T> findAll(Sort sort);
    Page<T> findAll(Pageable pageable);
}
```

这里的分页参数使用PageRequest类来构造。PageRequest是Spring提供的实现类，它实现了Java中的Pageable接口。下面是PageRequest提供的默认方法，已基本满足平常的业务需要。

```
public class PageRequest extends AbstractPageRequest {
    ...
    public static PageRequest of(int page, int size) {
        return of(page, size, Sort.unsorted());
    }
    public static PageRequest of(int page, int size, Sort sort) {
        return new PageRequest(page, size, sort);
    }
    ...
}
```

productRepository在声明时继承了JpaRepository类。JpaRepository继承了PagingAndSortingRepository类，在PagingAndSortingRepository类中声明了findAll方法。代码如下：

```
@NoRepositoryBean
public interface PagingAndSortingRepository<T, ID> extends CrudRepository<T, ID> {
    ...
    /**
     * Returns a {@link Page} of entities meeting the paging restriction provided
in the {@code Pageable} object.    *
     * @param pageable
     * @return a page of entities
     */
    Page<T> findAll(Pageable pageable);
}
```

返回的商品信息以及分页信息放入ModelAndView对象中，并将当前View映射到/home页面上，在前端/home页面中可以直接取出。

4.4.6 购物车与订单相关

1. 购物车展示

查询用户当前购物车内的所有商品列表。这个商品列表是存放在缓存中的，这里我们使用Map来模拟购物车缓存。

```
@GetMapping("/shoppingCart")
public ModelAndView shoppingCart() {
```

```java
        ModelAndView modelAndView = new ModelAndView("/shoppingCart");
        modelAndView.addObject("products", shoppingCartService.getProductsInCart());
        modelAndView.addObject("total", shoppingCartService.getTotal().toString());
        return modelAndView;
    }
```

getProductsInCart()方法获取的是商品列表。getTotal()获取的是购物车内所有商品的总价格。

```java
    public BigDecimal getTotal() {
        return products.entrySet().stream()
                .map(entry -> entry.getKey().getPrice().multiply(BigDecimal.valueOf(entry.getValue())))
                .reduce(BigDecimal::add)
                .orElse(BigDecimal.ZERO);
    }
```

获取购物车内所有商品总价格使用的是流方式。首先将Map转为流方式.stream()，然后将每个元素的价格乘以数量就是单个商品的总价，最后执行reduce方法，将所有元素相加得到总价格。调用完reduce方法会返回一个Optional<BigDecimal>对象，如果中间出现错误就无法得到正确的返回，从而影响系统使用。这里可以在reduce方法执行完之后加上一个orElse方法，设置一个出错之后的默认返回值。

2. 加入购物车

在商品展示页页，单击"购买"按钮，前端完成一系列验证后发送请求到后端，调用加入购物车接口，并在成功执行后将页面跳转到购物车页面，刷新购物车页面。

```java
    @GetMapping("/shoppingCart/addProduct/{productId}")
    public ModelAndView addProductToCart(@PathVariable("productId") Long productId) {
        productRepository.findById(productId).ifPresent(shoppingCartService::addProduct);
        return shoppingCart();
    }
```

这里使用了ifPresent方法，即如果findById方法有返回对象的话，执行后面的shoppingCartService中的addProduct方法，传入的参数即为findById返回的对象。

```java
    public void addProduct(Product product) {
        //如果购物车中存在这个商品，直接把数量加1
        if (products.containsKey(product)) {
            products.replace(product, products.get(product) + 1);
        } else {
            //如果购物车不存在这个商品，将商品放入购物车，数量设置为1
            products.put(product, 1);
        }
    }
```

3. 移除购物车

在购物车页面，单击"移除"按钮，发送移除请求到后端，调用移除接口，并在成功执行后将页面跳转到购物车页面，刷新购物车页面。移除时，同样先查询商品是否存在，若存在，就调用shoppingCartService中的removeProduct方法。

```java
public void removeProduct(Product product) {
    //如果购物车中存在这个商品
    if (products.containsKey(product)) {
        //如果数量大于1,则数量减1
        if (products.get(product) > 1)
            products.replace(product, products.get(product) - 1);
        else if (products.get(product) == 1) {
            //如果数量等于1,直接移除该商品
            products.remove(product);
        }
    }
}
```

根据商品信息判断商品数量,根据不同数量做不同操作。

4. 生成订单

后台接到生成订单请求后,首先获取当前登录用户信息,然后获取当前登录用户的购物车内容,生成对应订单。这里注意要对商品库存信息做校验。

```java
@GetMapping("/shoppingCart/checkout")
public ModelAndView checkout(Principal principal) {
    // 获取当前登录用户
    Optional<User> optionalUser = userRepository.findByUsername(principal.getName());
    if (!optionalUser.isPresent()) {
        throw new HttpClientErrorException(HttpStatus.UNAUTHORIZED);
    }
    try {
        // 生成订单
        shoppingCartService.checkout(optionalUser.get());
    } catch (NotEnoughProductsInStockException e) {
        // 商品库存不足
        return shoppingCart().addObject("outOfStockMessage", e.getMessage());
    }
    return shoppingCart();
}
```

商品库存信息不足采用的是手动抛出异常的方式来处理,该异常定义在exception包中。

```java
public class NotEnoughProductsInStockException extends Exception {
    private static final String DEFAULT_MESSAGE = "库存产品不足";
    public NotEnoughProductsInStockException() {
        super(DEFAULT_MESSAGE);
    }
    public NotEnoughProductsInStockException(Product product) {
        super(String.format("%s 库存不足。仅剩 %d 件。", product.getName(), product.getQuantity()));
    }
}
```

使用占位符的形式动态加载提示信息。这里还需要定义一个全局Handler类来处理所抛异常。

```java
@ExceptionHandler(Throwable.class)
@ResponseStatus(HttpStatus.INTERNAL_SERVER_ERROR)
public ModelAndView exception(final Throwable throwable, final Model model) {
    log.error("Exception during execution of SpringSecurity application", throwable);
    ModelAndView modelAndView = new ModelAndView("/error");
    String errorMessage = (throwable != null ? throwable.toString() : "Unknown error");
    modelAndView.addObject("errorMessage", errorMessage);
    return modelAndView;
}
```

统一将异常定向到error页面，在该页面提示错误信息。

生成订单用到的主要是checkout方法。

```java
public void checkout(User user) throws NotEnoughProductsInStockException {
    Product product;
    //生成订单基本信息：创建时间，对应用户
    Order order = new Order()
            .setCreateTime(LocalDateTime.now())
            .setUser(user);
    BigDecimal payment = BigDecimal.ZERO;
    //售卖记录列表
    List<Sold> soldList = new ArrayList<>();
    //遍历购物车内的商品
    for (Map.Entry<Product, Integer> entry : products.entrySet()) {
        Product key = entry.getKey();              //对应商品
        Integer quantity = entry.getValue();       //对应数量
        Optional<Product> one = productRepository.findOne(Example.of(key));// 判断商品是否存在
        if (!one.isPresent()) {
            throw new IllegalArgumentException("");
        }
        product = one.get();
        if (product.getQuantity() < quantity) {    //库存数量少于购物车中的数量
            throw new NotEnoughProductsInStockException(product);
        }
        //更新商品库存数量
        entry.getKey().setQuantity(product.getQuantity() - quantity);
        //新增商品售卖记录列表
        soldList.add(new Sold()
                .setQuantity(quantity)             //卖出数量
                .setProduct(key)                   //商品
                .setOrder(order));                 //订单
        payment = payment.add(key.getPrice());     //商品价格
    }
    order.setPayment(payment)                      //订单价格
            .setSoldList(soldList);                //售卖记录
    orderRepository.save(order);//保存订单，并根据Order实体类的设置级联保存售卖记录列表
    productRepository.saveAll(products.keySet()); // 更新产品库存信息
    productRepository.flush();                     //将当前数据库指令刷新到数据库
    products.clear();                              //完成之后，清除购物车
}
```

这里用到了前面所讲的级联操作。保存订单时需同步保存商品对应的售卖记录。对应的实体类配置如下：

```
// 订单实体类
public class Order implements Serializable {
    ...
    /*一对多，级联保存*/
    @OneToMany(fetch = FetchType.LAZY, mappedBy = "order",cascade = CascadeType.PERSIST)
    private List<Sold> soldList;
}
```

需要注意的是，以上操作涉及的表有多个，并且有多条SQL语句待执行，这种情况下需要借助事务来实现操作的完整性。对应的Service配置如下：

```
@Service
@Scope(value = WebApplicationContext.SCOPE_SESSION, proxyMode = ScopedProxyMode.TARGET_CLASS)
@Transactional
@RequiredArgsConstructor
public class ShoppingCartService{...}
```

@Transactional注解代表该Service下的方法默认使用事务。如果方法中抛出异常或出现其他错误，会自动执行回滚计划，避免出现脏数据。

@Scope注解用来指定Service的作用域。目前采用的是session作用域。每个作用域代表的意思如下：

- singleton单例模式(默认)：ConfigurableBeanFactory.SCOPE_SINGLETON，全局有且仅有一个实例。
- prototype原型模式：ConfigurableBeanFactory.SCOPE_PROTOTYPE，每次获取Bean的时候会有一个新的实例。
- request：WebApplicationContext.SCOPE_REQUEST，request表示针对每一次HTTP请求都会产生一个新的Bean，同时该Bean仅在当前HTTP request内有效。
- session：WebApplicationContext.SCOPE_SESSION，session作用域表示针对每一个会话请求都会产生一个新的Bean，同时该Bean仅在当前HTTP session内有效。

4.5 项目总结

本章实现了一个购物车管理系统开发，后端使用的主要技术有Spring、Spring MVC、Spring Data JPA和Spring Security，前端使用的技术主要有Thymeleaf，数据库使用的是H2。本章要掌握的重点是Spring Security、H2和Thymeleaf。知道如何通过Spring Boot整合这些框架并做相应配置，以及如何在实际开发过程中使用这些框架。学习时可参考各个框架官方提供的API手册，同时要回顾前面章节的相关内容，以便理解和记忆。

本章的重点技术：

- Spring Security框架：Spring Security的认证流程原理及使用方法，涉及的主要过滤器的作用。
- Thymeleaf框架：框架的引入方式、基本语法与规则，常用标签命令的使用方式和场景。
- H2数据库：H2数据库的下载与安装，各个模式的使用方式，在本项目中使用的模式。
- Bean Validation：Java Bean Validation验证规范以及常用的注解。
- Spring Data JPA级联操作：六种级联方式和使用方法。

第 5 章 用户权限管理系统

在第4章中,我们介绍并使用Spring Security、H2数据库以及Thymeleaf技术完成了商城购物车功能的开发与实现。本章将实现一个用户权限管理系统,包括账号管理、菜单和按钮权限控制、角色及角色类型、组织管理等内容。在技术实现方面,后台采用Spring Boot、MyBatis,安全登录框架为Shrio权限控制框架,前端页面继续使用第4章介绍的Thymeleaf框架和第2章使用的LayUI框架,采用的数据库是PostgreSQL数据库。

本章主要涉及的知识点有:

- 如何使用Spring Boot集成Shrio框架。
- 如何使用Spring Boot集成PostgreSQL。
- 如何使用Shrio安全框架。
- 如何使用PostgreSQL进行数据的增、删、改、查。

5.1 项目技术选型

本章的用户权限管理系统涉及的功能,是很多系统都要使用的基础权限控制功能,根据项目的规模可以单独部署,也可以集成到项目中去。业务范围比较简单,主要讲解所用框架技术,同时对项目模块进行了划分,整个项目未采取前后端分离技术。业务代码选用的框架是Spring MVC、MyBatis,权限认证框架采用Shrio,数据库使用PostgreSQL,页面选用的是Thymeleaf模板技术,布局采用的是LayUI。

5.1.1 Shrio权限认证框架

1. 概述

Apache Shiro是一个功能强大、灵活易用的Java开源安全框架,可以非常优雅清楚地处理身份验证、授权、企业会话管理和加密。

ApacheShiro的初衷及终极目标是让使用者易于使用和理解。确实，安全性有时可能非常复杂，甚至令开发者苦不堪言，但不一定都是如此。一个优秀的安全框架应该尽可能地掩盖复杂性，并公开一个简洁、直观、可读性强的API，以简化开发人员在应用程序安全方面的工作。

目前Shiro作为一个企业应用程序，并不强制依赖其他第三方框架、容器或应用程序服务器，它尽可能地融入这些环境，以实现在任何环境中开箱即用。

2. Apache Shiro特性

Apache Shiro是一个具有许多功能的综合应用程序安全框架，整体功能大致如图5.1所示。

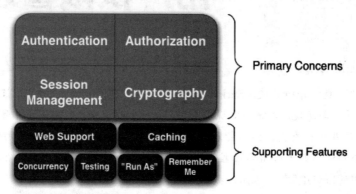

图5.1　Apache Shrio整体功能图

以上每一个色块都代表一个功能模块。

- Authentication：身份认证/登录，验证用户是不是合法用户。
- Authorization：授权，即权限验证，验证某个已认证的用户是否拥有某个权限。例如，验证某个用户是否拥有某个角色，或者验证某个用户对某个菜单是否具有某个权限。
- Session Management：会话管理，即用户登录后就是一次会话，在没有退出之前，它的所有信息都在会话中。
- Cryptography：加密，保护数据的安全性，例如将密码加密存储到数据库，而不是直接存储明文。
- Web Support：Web支持，集成到Web环境简便快捷。
- Caching：缓存，比如，用户登录后缓存用户信息、拥有的角色/权限，提高性能和效率。
- Concurrency：Shiro支持多线程应用的并发验证，即如果在一个线程中开启另一个线程，能实现权限共享。
- Testing：提供测试支持。
- Run As：在获得允许的前提下，允许一个用户假装为另一个用户进行访问。
- Remember Me：记住我，记住用户信息，只有在必须时才需要重新登录。

3. Apache Shiro架构

首先，我们从整体的角度来看一下Shrio框架的组成。Shrio框架有3个主要的概念：Subject、SecurityManager和Realms。图5.2展示了这些组件之间是如何交互的。

第 5 章 用户权限管理系统

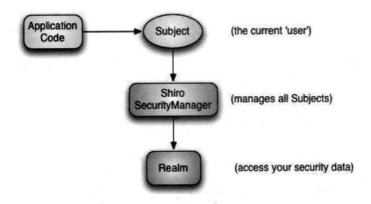

图5.2　Shrio框架图

- Subject：主体，代表了当前"用户"，这个用户不一定是一个具体的人，与当前应用交互的任何内容都是Subject，例如第三方服务、后台账户或其他类似的事物。所有Subject都会且必须绑定到SecurityManager，与Subject的所有交互都由SecurityManager托管，Subject只是一个接收者，SecurityManager是实际执行者。
- SecurityManager：安全管理器，它是Shiro的核心。它协调内部的安全组件，将这些组件组合到一起形成一个对象图。
- Realm：域，它像一座桥或者一个连接器，将Shrio和当前应用连接起来。当进行安全数据交互时，例如使用用户数据进行登录或权限控制，Shiro从Realm中获取安全数据。在这种模式下，Realm 可以看作一个数据源。配置Shrio时，SecurityManager支持多个Realm，但至少要指定一个Realm。我们一般在应用中都需要实现自己的 Realm。

以上内容概括来说，就是应用通过Subject来进行认证和授权，而Subject实际上又委托给了SecurityManager，所以配置Shrio时，需要给SecurityManager 注入至少一个Realm，从而能让SecurityManager得到合法的用户安全数据及其权限进行校验。

看完整体架构后，接下来看一下详细的架构。这里主要分为三部分，如图5.3所示。最上层为要集成Shrio的应用代码，它支持多种语言，可以是C/C++、Python、Flex、Web应用等。中间层是Shrio核心架构内容，也是接下来重点讲述的部分。底层是数据源。

详细架构部分主要描述了Security Manager和cryptography两部分，其中Security Manager部分是Shrio框架的核心，所有具体的交互都通过SecurityManager进行控制，它管理着所有的Subject，并且负责进行认证、授权、会话、缓存的管理。SecurityManager具体包含了以下几部分内容：

- Authenticator(org.apache.shiro.authc.Authenticator)：认证器，负责用户登录认证。认证器从Realm中获取用户信息用于认证。
- AuthenticationStrategy(org.apache.shiro.authc.pam.AuthenticationStrategy)：认证策略，如果配置了多个Realm，则由认证策略负责决定使用哪个Realm。
- Authorizer(org.apache.shiro.authz.Authorizer)：访问控制器，用来决定用户是否有权限进行相应的操作。

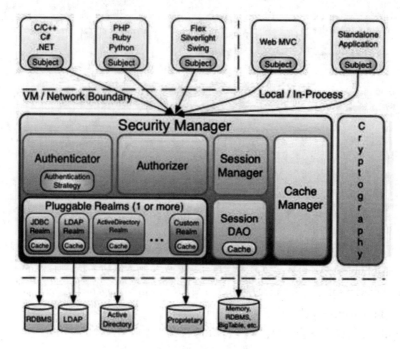

图5.3　Shrio架构设计

- SessionManager(org.apache.shiro.session.mgt.SessionManager)：负责管理用户Session的生命周期。Shrio能够在任何环境下管理内部的用户session，不管是不是Web应用，对Shrio来说都不会有影响。如果应用中存在session的话，比如Servler容器，则Shrio会利用现存的session机制。如果没有的话，比如非Web应用，就会基于自己的session机制。
- SessionDAO(org.apache.shiro.session.mgt.eis.SessionDAO)：数据访问对象，处理Session到数据库的持久化操作。
- CacheManager(org.apache.shiro.cache.CacheManager)：缓存控制器，用于管理缓存，提高访问性能。
- Cryptography(org.apache.shiro.crypto.*)：加密模块，Shiro提供了一些易用的、易于理解的、常见的加密组件用于密码加密/解密。

4．身份验证

身份验证是指在应用中证明用户即为所声称的身份，一般需要提供一些证据信息来表明。在Shrio中，用户需要提供 principals（身份）和credentials（证明）给Shiro，从而验证用户身份。

- principals：身份，即主体的标识属性，可以是任何东西，如账号、邮箱、手机号等。一个主体可以有多个 principals，但只有一个 Primary principals，一般是用户名、邮箱或用户在应用中的全局ID。
- credentials：证明/凭证，即只有主体知道的安全值，如密码、手印、视网膜扫描信息等。

最常见的principals和credentials组合就是用户名和密码了。

使用Shrio进行简单的身份认证，基本步骤分为三步：

步骤 01　收集用户身份/凭证。
步骤 02　提交信息用于认证。
步骤 03　提交返回结果成功，认证通过；否则重新认证或者阻止访问。

接下来我们对对应代码进行分析。

第一步，收集用户身份和凭证信息。

```
//使用最常见的用户名和密码
UsernamePasswordToken token = new UsernamePasswordToken(username, password);
//开启记住我
token.setRememberMe(true);
```

UsernamePasswordToken即用户名密码Token，实现了AuthenticationToken接口。

第二步，提交用户身份和凭证，调用Subject.login进行登录。

```
Subject currentUser = SecurityUtils.getSubject();
try {
    currentUser.login(token);
} catch ( UnknownAccountException uae ) {
    ... //账号不存在
} catch ( IncorrectCredentialsException ice ) {
    ... //密码错误
} catch ( LockedAccountException lae ) {
    ...//账户状态有误
} catch ( ExcessiveAttemptsException eae ) {
    ... //超过错误次数
} catch ( AuthenticationException ae ) {
    ...//其他错误
}
```

第三步，根据login方法的执行结果返回对应的结果。如果失败将得到相应的异常，根据异常提示用户错误信息，否则登录成功；Subject收到login方法请求后，会自动委托给SecurityManager，而SecurityManager会委托给Authenticator进行身份验证，Authenticator可能会委托给相应的AuthenticationStrategy进行多Realm身份验证，然后Authenticator会把相应的token传入Realm，从Realm获取身份验证信息，如果没有返回并且没有抛出异常表示身份验证成功了。

注销时，调用logout方法。

```
currentUser.logout();
```

以上的使用过程如图5.4所示。

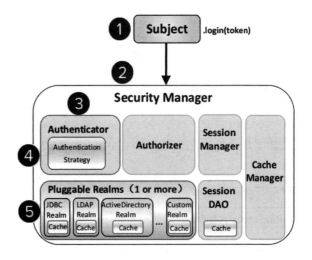

图5.4　身份验证过程

5. Shiro授权

授权也叫访问控制，即在应用中控制谁能访问哪些资源，例如访问页面、编辑数据、按钮操作、页面打印等。这里涉及几个对象：权限（Permissions）、角色（Roles）、用户（Users）。

（1）基本对象

- 权限（Permissions）

权限代表了在应用中可以做的操作。一个完整的权限包含所涉及的资源以及可以对某项资源所做的操作。大多数资源都包含增、删、改、查的行为操作。所以，权限是基于资源（Resources）和操作（Actions）的。资源有很多种，例如URL、页面、数据、方法等都是资源。用户只有被授权后才能进行相应的操作。

- 角色（Roles）

角色代表了操作集合，可以理解为权限的集合，一般情况下我们会被赋予用户角色而不是权限，即这样用户可以拥有该角色下的权限集合。Shrio中有两种角色，隐式角色和显式角色。隐式角色直接通过角色来验证用户有没有操作权限，例如，代码约定角色admin默认拥有用户管理权限，仅通过角色的名字即可决定权限，而不去考虑这个角色下是否绑定了用户管理权限。这种形式的可维护性不高，如果要更改需求，需要重新修改代码。显示角色在程序中通过权限控制谁能访问某个资源，角色下绑定一组权限集合，可以动态地增加和删除权限，这种形式的可维护性高。

（2）授权方式

Shiro支持三种方式的授权：

- 编程式：通过代码块完成，例如 if...else 授权代码块。
- JDK注解式：通过在方法上添加相应的注解完成。
- JSP/GSP标签：通过页面上相应的角色权限标签完成。

① 编程式

通过代码块实现授权时，可以使用Subject中的角色方法hasRole，代码如下：

```
Subject subject = SecurityUtils.getSubject();
if(subject.hasRole("admin")) {
    //有权限
} else {
    //无权限
}
```

此类角色方法在Subject中有3个，如表5.1所示。

表5.1　Subject类中的角色方法

方 法 名	描　述
hasRole(String roleName)	拥有角色则返回true，否则返回false
hasRoles(List<String> roleNames)	返回一个包含true或false的数组，每个元素和单独调用hasRole的返回结果相同
hasAllRoles(Collection<String> roleNames)	拥有列表中的所有角色，则返回true，否则返回false

也可以通过角色校验方法checkRole实现授权，代码如下：

```
Subject currentUser = SecurityUtils.getSubject();
//guarantee that the current user is a bank teller and
//therefore allowed to open the account:
currentUser.checkRole("bankTeller");
openBankAccount();
```

如果有权限，则继续向下运行，如果没有权限则抛出AuthorizationException异常。该异常可自定义。

角色校验的方法在Subject中同样也有3个，如表5.2所示。

表5.2　Subject中的角色校验方法

方 法 名	描 述
checkRole(String roleName)	如果没有当前角色授权，则抛出异常，否则无动作
checkRoles(Collection<String> roleNames)	如果不满足当前所有角色的授权，则抛出异常，否则无动作
checkRoles(String… roleIdentifiers)	与checkRoles(Collection<String> roleNames)规则相同

不仅可以校验角色，还可以校验权限，代码如下：

```
Permission printPermission = new PrinterPermission("laserjet4400n", "print");
Subject currentUser = SecurityUtils.getSubject();
if (currentUser.isPermitted(printPermission)) {
    //显示打印标签
} else {
    //没有授权的处理
}
```

Subject中通过权限对象进行校验的权限判断方法如表5.3所示。

表5.3　Subject中的权限判断方法

方 法 名	描 述
isPermitted(Permission p)	拥有权限，则返回true，否则返回false
isPermitted(List<Permission> perms)	将集合中每一个权限的校验结果组合成一个布尔数组返回
isPermittedAll(Collection<Permission> perms)	拥有列表中的所有权限，则返回true，否则返回false

也可以通过字符串进行校验，代码如下：

```
Subject currentUser = SecurityUtils.getSubject();
if (currentUser.isPermitted("printer:print:laserjet4400n")) {
   ...
} else {
   ...
}
```

Subject中通过字符串进行校验的权限判断方法如表5.4所示。

表5.4 Subject中的权限判断方法

方法名	描述
isPermitted(String perm)	拥有字符串中的权限,则返回true,否则返回false
isPermitted(List perms)	根据列表中的权限返回一个布尔数组
isPermittedAll(String… perms)	拥有列表中的所有权限,则返回true,否则返回false

权限校验同样也支持抛出异常的形式。直接调用checkPermission方法,如果未通过会抛出异常,通过后就会继续向下执行。代码如下:

```
Subject currentUser = SecurityUtils.getSubject();
//guarantee that the current user is permitted
//to open a bank account:
currentUser.checkPermission("account:open");
openBankAccount();
```

这个方法同时支持前面提到的权限对象和字符串。

② JDK注解式

通过注解实现授权时,可以直接在方法上添加授权注解,代码如下:

```
@RequiresRoles("admin")
public void hello() {
    //有权限
}
```

当满足注解所设置的权限条件时,可执行该方法,如果不满足,则抛出AuthorizationException异常,这些异常可以根据情况自定义。

常用的注解及作用说明如表5.5所示。

表5.5 权限校验注解

名称	作用
@RequiresAuthentication	验证用户是否登录,等同于subject.isAuthenticated()方法的结果为true时
@RequiresGuest	验证未登录用户或者未被之前的session记住,与@RequiresUser完全相反
@RequiresPermissions("account:create")	需要account下的create权限
@RequiresRoles("admin")	需要拥有admin角色
@RequiresUser	需要用户对象,可以是成功登录的,也可以是之前被记住的,与@RequiresGuest完全相反

③ JSP/GSP标签

通过页面标签进行授权,可以使用现有的标签库。通过如下语句导入标签:

```
<%@ taglib prefix="shiro" uri="http://shiro.apache.org/tags" %>
```

然后在页面代码块外层包裹如下代码:

```
<shiro:hasRole name="admin">
<!-- 有权限 -->
</shiro:hasRole>
```

标签支持的命令和参数与上述通过代码方式授权基本相同,可以参考详细的官方API文档。这里不再赘述。

5.1.2 PostgreSQL数据库

1. 概述

PostgreSQL 是开源的对象关系数据库管理系统（ORDBMS），它是以加州大学伯克利分校计算机系开发的POSTGRES 4.2版本为基础开发的。它支持大部分SQL标准,有以下特性：

- 复杂查询。
- 外键。
- 触发器。
- 视图。
- 事务完整性。
- 多版本并发控制。

它也支持用户使用自定义方法进行扩展,比如可以增加以下内容：

- 数据类型。
- 函数。
- 操作符。
- 聚集函数。
- 索引方法。
- 过程语言。

它是一个开源项目,任何人都可以拿到它的源码进行使用、修改和分发,不论使用目的是自用、商业用途还是学术研究。

2. PostgreSQL简史

PostgreSQL发展到目前的规模,经历了不同时代的版本迭代。由最初的概念具象到实际的应用,再由最初的应用逐渐发展成目前较为成熟的版本,每一次版本更迭都是它的一大进步。它最初始于1986年,当时提出了该系统最初的概念、最早的数据模型定义、最早的规则系统设计以及存储管理器的理论基础和体系结构。从那以后,POSTGRES经历了几次主要的版本更新。第一个"演示性"系统在1987年便可使用了,并且在1988年的ACM SIGMOD大会上展出。在1989年6月发布了版本1。为了回应用户对第一个规则系统的改进意见,在1990年6月发布了使用新规则系统的版本2。版本3在1991年出现,增加了多存储管理器的支持,并且改进了查询执行器,重新编写了规则系统。从那以后的版本直到Postgres95发布前,其主要工作都集中在移植性和可靠性上。到了1993年,外部用户的数量几乎翻倍,导致用于源码维护的时间日益增加,以至占用了太多本应该用于数据库研究的时间,为了减少负担,伯克利的POSTGRES项目在版本4.2时正式终止。1994年，Andrew Yu和Jolly Chen向POSTGRES中增加了SQL语言的解释器，并将Postgres95的源码公开供大家使用。到了1996年，一个新名字PostgreSQL出现了，这个名字能够反映最初的POSTGRES和最新的使用SQL的版本之间的关系，同时版本号也重新从6.0开始，将版本号回溯到最初由伯克利开发的POSTGRES项目的版本顺序中。从此以后，PostgreSQL发行的版本都可以在官方文档中追溯。整个PostgreSQL发展史如图5.5所示。

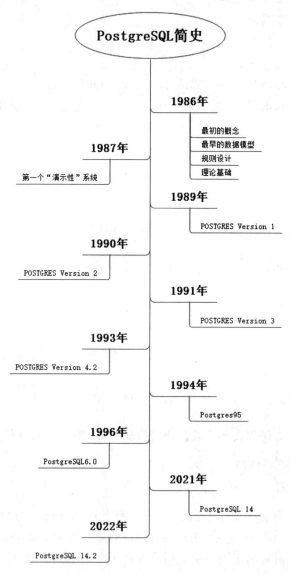

图5.5　PostgreSQL简史

3. 下载和安装

PostgreSQL官网地址为https://www.postgresql.org/，下载地址为https://www.enterprisedb.com/downloads/postgres-postgresql-downloads。这里以Windows系统为例，演示它的下载和安装过程。

步骤01　访问上述下载地址，下载对应的安装包，如图5.6所示。

第 5 章 用户权限管理系统 | 181

图5.6 PostgreSQL下载页面

步骤 02 双击下载的安装包，弹出欢迎对话框，如图5.7所示。
步骤 03 在欢迎对话框中单击"Next"按钮进入下一步，选择安装位置，这里根据自身情况自定义即可，如图5.8所示。

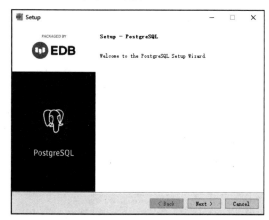

图5.7 PostgreSQL欢迎对话框

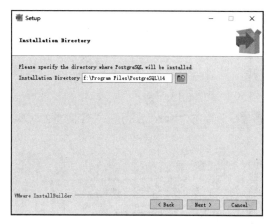

图5.8 选择安装位置

步骤 04 单击"Next"按钮进入下一步，选择默认配置，如图5.9所示。
步骤 05 单击"Next"按钮进入下一步，选择数据文件存放地址，默认为安装目录下的data文件，这里可以不做更改，如图5.10所示。

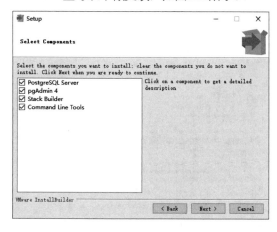

图5.9 选择默认配置

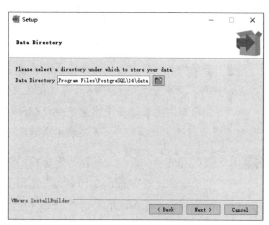

图5.10 数据存放目录设置

步骤 06 单击"Next"按钮进入下一步，设置超级管理员的初始密码，如图5.11所示。

步骤 07　单击"Next"按钮进入下一步，设置数据库服务的端口，默认是5432，如果和现存服务的端口存在冲突，则需要更改，另外选择一个闲置端口即可，如图5.12所示。

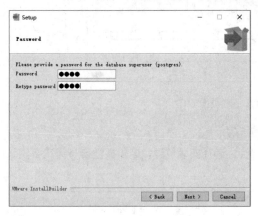
图5.11　设置管理员和密码

图5.12　设置端口

步骤 08　单击"Next"按钮进入本地化语言选择页面，单击"Next"按钮进入下一步，确认安装信息，如图5.13所示。

步骤 09　单击"Next"按钮进入安装过程，如图5.14所示。该过程较长，需耐心等待。

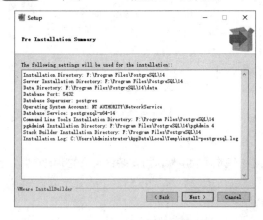
图5.13　确认安装信息

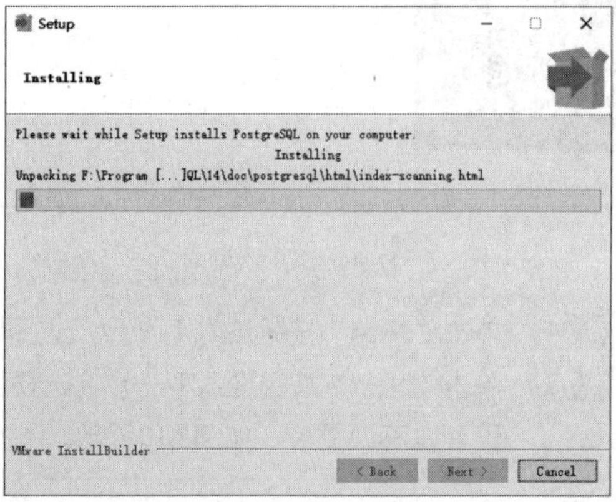

图5.14　安装进程

注意：如果安装过程中运行防火墙或者其他安全保护软件，可能会因为权限问题造成安装失败，如图5.15所示。安装过程中注意及时处理授权弹窗或者暂时退出此类软件，等安装完成后再运行。

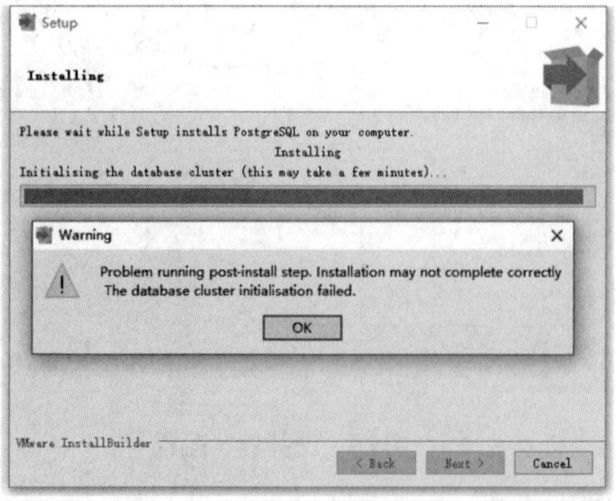

步骤 10　安装完成后，取消勾选运行复选框，关闭安装窗口。在Windows 10桌面左下角的搜索框内输入pgAdmin命令，如图5.16所示。

步骤 11　进入pgAdmin 4可视化管理客户端后，弹窗提示输入密码，这里输入刚才设置的超级管理员密码即可，如图5.17所示。

图5.15　安装失败示例

 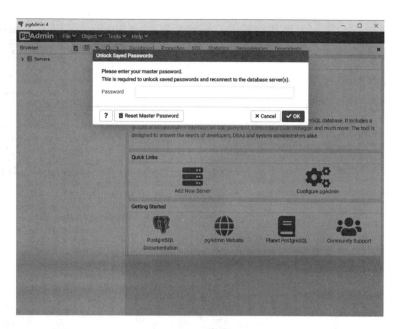

图5.16　客户端打开方式　　　　　　　　图5.17　客户端使用

进入管理页面后，可以看到当前数据库的监控页面，如图5.18所示。

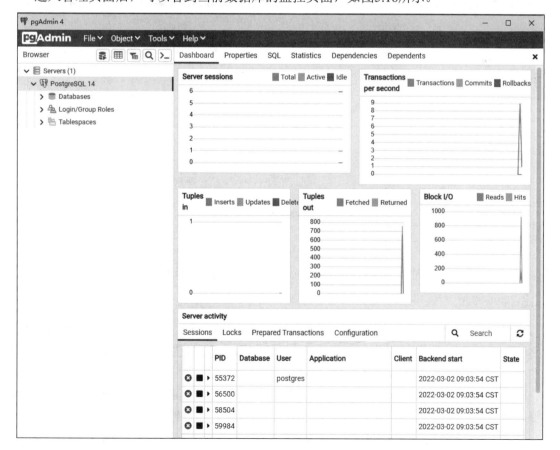

图5.18　客户端主页面

在这里可以操作数据库。当然也可以选择用命令行操作数据库。

步骤12 在计算机搜索中查找SQL Shell(psql)，如图5.19所示。

步骤13 进入SQL Shell(psql)后，Server、Database、Port、Username这四项内容无须输入，直接按回车键即可，在用户"postgres"的口令中输入管理员密码，如图5.20所示。

如果上述四项内容在安装过程中有改动，即与后面中括号内提示的内容不同时，需要输入自定义的内容方可进入。

进入之后可以执行SQL命令，如遇到不清楚的命令，可以使用help命令来提示，如图5.21所示。

图5.19　命令行打开方式

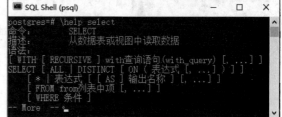

图5.20　Shell窗口　　　　　　　　　　图5.21　命令示例

这个命令行窗口也可以通过直接打开文件的形式打开。它的运行文件默认存放在安装目录下的script目录中，如图5.22所示。

图5.22　命令行运行文件

4．语法

PostgreSQL支持标准的SQL语法，由于篇幅关系这里不再赘述。

5．数据类型

PostgreSQL支持多种数据类型，同时支持用户使用CREATE TYPE命令添加新的数据类型，定义该类型的用户成为其所有者。如果给出模式名，那么该类型是在指定模式中创建，否则它将在当前模式中创建。类型名必需和同一模式中任何现有的类型或者域不同。因为表和数据类型有联系，所以类型名也不能和同模式中的表名字冲突。可以创建复合类型，也可以创建基本类型。创建复合类型时，语法相对简单，是通过一列属性名和数据类型声明的。创建基本类型时，相对复杂，需要定义input_function和output_function函数，这里就涉及C或者其他底层语言了，所以不再详述。语法示例如下：

```
--复合类型
CREATE TYPE name AS
    ( attribute_name data_type [, ... ] )
--基本类型
CREATE TYPE name (
    INPUT = input_function,
    OUTPUT = output_function
    [ , RECEIVE = receive_function ]
    [ , SEND = send_function ]
    [ , ANALYZE = analyze_function ]
    [ , INTERNALLENGTH = { internallength | VARIABLE } ]
    [ , PASSEDBYVALUE ]
    [ , ALIGNMENT = alignment ]
    [ , STORAGE = storage ]
    [ , DEFAULT = default ]
    [ , ELEMENT = element ]
    [ , DELIMITER = delimiter ]
)
```

复合类型可以理解为某些常用字段组合在一起作为一个变量同时出现。例如经纬度就可以定义一个复合类型，由两个float类型或者decimal类型组成。

接下来看一下PostgreSQL支持的数据类型。

（1）数值类型

PostgreSQL中的数值类型如表5.6所示。其中，smallint、integer和bigint是整数类型；serial、smallserial和bigserial是整数中的自增正数，不存负数。

表5.6　PostgreSQL中的数值类型

名　　字	存储空间	描　　述	范　　围
smallint	2 字节	小范围的整数	−32768～+32767
integer	4 字节	常用的整数	−2147483648～+2147483647
bigint	8 字节	大范围的整数	−9223372036854775808～9223372036854775807
decimal	可变长度	用户声明精度，精确	无限制
numeric	可变长度	用户声明精度，精确	无限制
real	4 字节	变精度，不精确	6 位十进制数字精度
double precision	8 字节	变精度，不精确	15 位十进制数字精度
serial	4 字节	自增整数	1～2147483647
smallserial	2 字节	自增整数	1～32767
bigserial	8 字节	大范围的自增整数	1～9223372036854775807

（2）货币类型

PostgreSQL中的货币类型如表5.7所示。money类型存储带有固定小数精度的货币金额。numeric、int和bigint类型的值可以转换为 money，不建议使用浮点数来处理处理货币类型，因为可能会丢失精度。精度可以通过lc_monetary变量配置。

表5.7　PostgreSQL中的货币类型

名　字	存储空间	描　述	范　围
money	8字节	货币金额	−92233720368547758.08～+92233720368547758.07

（3）字符类型

PostgreSQL中的字符类型在PostgreSQL里可用于一般用途的字符类型，如表5.8所示。

表5.8　PostgreSQL中的字符类型

名　字	描　述
character varying(n), varchar(n)	变长，有长度限制
character(n), char(n)	定长，不足补空白
Text	变长，无长度限制

SQL定义了两种基本的字符类型：character varying(n)和character(n)，这里的n是一个正整数。两种类型都可以存储最多n个字符的字符串。如果要存储的字符串比声明的长度短，类型为character的数值将会用空白填满；而类型为character varying的数值将会存储实际的字符串。varchar(n)和char(n)分别是character varying(n)和character(n)的别名，没有声明长度的character等于character(1)；如果使用character varying时不带长度，那么该类型接收任何长度的字符串。

另外，PostgreSQL提供text类型，它可以存储任何长度的字符串。尽管text类型不是SQL标准，但是许多其他SQL数据库系统也有它。

注意：这三种类型之间没有性能差别，只不过是在使用character的时候增加了存储尺寸。虽然在某些数据库系统里，character(n)有一定的性能优势，但在PostgreSQL里没有。在大多数情况下，应该使用text或character varying。

在PostgreSQL里另外还有两种定长字符类型，如表5.9所示。name类型只用于在内部系统表中存储标识符并且不是给一般用户使用的。该类型长度当前定为64字节（63字节可用字符加1字节结束符），但应该使用常量NAMEDATALEN引用。这个长度是在编译的时候设置的，因而可以为特殊用途调整，默认的最大长度在以后的版本可能会改变。类型"char"（注意引号）和char(1)是不一样的，它只占用1字节的存储空间。它在系统内部作为枚举类型用于系统表中。

表5.9　PostgreSQL中的定长字符类型

名　字	存储空间	描　述
"char"	1 字节	单字节内部类型
name	64 字节	用于对象名的内部类型

（4）二进制数据类型

PostgreSQL中的二进制数据类型如表5.10所示。

表5.10　PostgreSQL中的二进制数据类型

名　字	存储空间	描　述
bytea	4 字节加上实际的二进制字符串	变长的二进制字符串

二进制字符串是一个字节序列。在一条SQL语句的文本串里面输入bytea数值的时候，必须对某些字节值进行转义。通常，需要把它的数值转换成对应的3位八进制数，并且前导两个反斜杠。表5.11列出了必须转义的序列并给出了可选的转义序列。

表5.11 转义序列

十进制数值	描述	输入转义形式	例子	输出形式
0	八进制的零	E'\\000'	SELECT E'\\000'::bytea;	\000
39	单引号	'''' 或 E'\\047'	SELECT E'\''::bytea;	'
92	反斜杠	E'\\\\' 或 E'\\134'	SELECT E'\\\\'::bytea;	\\
0～31以及127～255	"不可打印"字节	E'\\xxx'(八进制值)	SELECT E'\\001'::bytea;	\001

bytea字节在输出中逃逸。通常，每个"不可打印"的字节值都转化成对应的前导反斜杠的三位八进制数值。大多数"可打印"的字节值以客户端字符集的标准表现形式出现。十进制值为92（反斜杠）的字节有一个特殊的可选输出形式。bytea输出转义序列如表5.12所示。

表5.12 bytea输出转义序列

字节的十进制值	描述	逃逸的转义形式	例子	输出结果
92	反斜杠	\\	SELECT E'\\134'::bytea;	\\
0～31以及127～255	"不可打印"八进制字符	\xxx(八进制值)	SELECT E'\\001'::bytea;	\001
32～126	"可打印"八进制字符	客户端字符集表现形式	SELECT E'\\176'::bytea;	~

SQL标准定义了一种不同的二进制字符串类型，例如BLOB或BINARY LARGE OBJECT，其输入格式和bytea不同，但提供的函数和操作符基本一样。

（5）时间/日期类型

PostgreSQL支持SQL中所有日期和时间类型，如表5.13所示。

表5.13 PostgreSQL中的日期和时间类型

名字	存储空间	描述	最低值	最高值	分辨率
timestamp [(p)] [without time zone]	8字节	日期和时间	4713 BC	5874897 AD	1毫秒/14位
timestamp [(p)] with time zone	8字节	日期和时间，带时区	4713 BC	5874897 AD	1毫秒/14位
interval [(p)]	12字节	时间间隔	−178000000年	178000000年	1毫秒/14位
date	4字节	只用于日期	4713 BC	5874897 AD	1天
time [(p)] [without time zone]	8字节	只用于一日内时间	00:00:00	24:00:00	1毫秒/14位
time [(p)] with time zone	12字节	只用于一日内时间，带时区	00:00:00+1459	24:00:00-1459	1毫秒/14位

对于time类型，如果使用8字节的整数存储，那么p允许的范围是0~6，如果使用的是浮点数存储，那么这个范围是0~10。

time with time zone类型是SQL标准定义的。在大多数情况下，表5.13所列的数据类型组合就能提供一切应用需要的日期/时间的完整功能。

date类型的输入方式如表5.14所示。

表5.14 日期输入

例子	描述
January 8, 1999	在任何DateStyle输入模式下都无歧义
1999-01-08	ISO 8601格式（建议格式），任何方式下都是1999年1月8号
1/8/1999	有歧义，在MDY模式下是一月八日；在DMY模式下是八月一日
1/18/1999	在MDY模式下是一月十八日，其他模式下被拒绝
01/02/03	在MDY模式下是2003年1月2日；在DMY模式下是2003年2月1日；在YMD模式下是2001年2月3日
1999-Jan-08	任何模式下都是1月8日
Jan-08-1999	任何模式下都是1月8日
08-Jan-1999	任何模式下都是1月8日
99-Jan-08	在YMD模式下是1月8日，否则错误
08-Jan-99	一月八日，除了在YMD模式下是错误的之外
Jan-08-99	一月八日，除了在YMD模式下是错误的之外
19990108	ISO 8601格式，任何模式下都是1999年1月8日
990108	ISO 8601格式，任何模式下都是1999年1月8日
1999.008	年和年里的第几天
J2451187	儒略日
January 8, 99 BC	公元前99年

当日时间类型是time without time zone、time with time zone。只写time等效于time without time zone。

这些类型的有效输入由当日时间后面跟着可选的时区组成（参见表5.15和表5.16）。如果在time without time zone类型的输入中声明了时区，那么它会被悄悄地忽略。同样指定的日期也会被忽略，除非使用了一个包括夏令时规则的时区名，比如America/New_York，在这种情况下，必须指定日期以确定这个时间是标准时间还是夏令时。时区偏移将记录在time with time zone中。

表5.15 时间输入

例子	描述
04:05:06.789	ISO 8601
04:05:06	ISO 8601
04:05	ISO 8601
040506	ISO 8601
04:05 AM	与04:05一样；AM不影响数值

(续表)

例　子	描　述
04:05 PM	与 16:05 一样；输入小时数必须≤12
04:05:06.789-8	ISO 8601
04:05:06-08:00	ISO 8601
04:05-08:00	ISO 8601
040506-08	ISO 8601
04:05:06 PST	缩写的时区
2003-04-12 04:05:06 America/New_York	用名字声明的时区

表5.16　时区输入

例　子	描　述
PST	太平洋标准时间（Pacific Standard Time）
America/New_York	完整时区名称
PST8PDT	POSIX风格的时区
-8:00	ISO-8601与PST的偏移
-800	ISO-8601与PST的偏移
-8	ISO-8601与PST的偏移
zulu	军方对UTC的缩写（译注：可能是美军）
z	zulu 的缩写

时间戳类型的有效输入由一个日期和时间的连接组成，后面跟着一个可选的时区，一个可选的AD（公元后）或BC（公元前）。另外，AD/BC可以出现在时区前面，但这个顺序并非最佳的。因此1999-01-08 04:05:06和1999-01-08 04:05:06 -8:00都是有效的数值，它是兼容ISO 8601的。另外，也支持下面这种使用广泛的格式：

```
January 8 04:05:06 1999 PST
```

时间间隔interval

interval数值可以用下面的语法声明：

```
[@] quantity unit [quantity unit...] [direction]
```

其中，quantity是一个数字（可能有符号）；unit是second、minute、hour、day、week、month、year、decade、century、millennium或者这些单位的缩写或复数；direction可以是ago或者为空；@符号是可选的。不同的单位以及相应正确的符号都是隐含地增加的。

- 特殊值

为方便起见，PostgreSQL支持表5.17中显示的几个特殊输入值。infinity和-infinity值是在系统内部显示的，并且将按照同样的方式显示；其他的特殊值都只是符号缩写，在读取的时候将被转换成普通的日期/时间值。特别是now和其相关的字符串在读取的时候就被转换成对应的数值。所有这些值在SQL命令里被当作普通常量对待时，都需要写在单引号里面。

（6）布尔型

PostgreSQL支持SQL标准的布尔数据类型。布尔值只能有"true"（真）或"false"（假），"unknown"（未知）状态用NULL表示。

表5.17 特殊日期/时间输入

输入字符串	适用类型	描述
epoch	date, timestamp	1970-01-01 00:00:00+00（Unix系统零时）
infinity	timestamp	比任何其他时间戳都晚
-infinity	timestamp	比任何其他时间戳都早
now	date, time, timestamp	当前事务的开始时间
today	date, timestamp	今日午夜
tomorrow	date, timestamp	明日午夜
yesterday	date, timestamp	昨日午夜
allballs	time	00:00:00.00 UTC

"真""假"值对应的有效文本值如表5.18所示。

表5.18 布尔型

真　　假	有效文本值
真	TRUE
	't'
	'true'
	'y'
	'yes'
	'1'
假	FALSE
	'f'
	'false'
	'n'
	'no'
	'0'

（7）枚举类型

枚举类型包含一组静态的按顺序排列的集合值，与编程语言中的enum类型基本相同。常见的有一周中每一天的称呼或某一数据项的状态值。

可以通过CREATE TYPE命令创建枚举类型。例如，创建一周中每天的称呼：

```
CREATE TYPE weekday AS ENUM ('Monday', 'Tuesday','Wednesday','Thursday','Friday',
'Saturday','Sunday');
```

枚举类型中各个值的排列顺序以声明时的顺序为准。如果针对枚举类型进行排序或者比较大小，会根据声明的排序进行。例如，一周中每天对应不同的课程，首先向class表中插入数据：

```
CREATE TABLE class( name text, display_weekday weekday);
INSERT INTO class VALUES ('语文', 'Monday');
INSERT INTO class VALUES ('数学', 'Tuesday');
INSERT INTO class VALUES ('体育', 'Friday');
```

若想查询周三以后的课程列表，代码如下：

```
SELECT * FROM class WHERE display_weekday > 'Wednesday';
```

若想按照周几排列课程，代码如下：

```
SELECT * FROM class  ORDER BY display_weekday;
```

（8）几何类型

几何类型表示二维的平面物体。PostgreSQL中可用的几何类型如表5.19所示。最基本的几何类型是点，是其他类型的基础。

表5.19　几何类型

名　字	存储空间	说　　明	表现形式
point	16字节	平面中的点	(x,y)
line	32字节	（无穷）直线（未完全实现）	((x1,y1),(x2,y2))
lseg	32字节	（有限）线段	((x1,y1),(x2,y2))
box	32字节	矩形	((x1,y1),(x2,y2))
path	16+16n字节	闭合路径（与多边形类似）	((x1,y1),…)
path	16+16n字节	开放路径	[(x1,y1),…]
polygon	40+16n字节	多边形（与闭合路径相似）	((x1,y1),…)
circle	24字节	圆	<(x,y),r>（圆心和半径）

（9）网络地址类型

PostgreSQL提供用于存储IPv4、IPv6、MAC地址的数据类型，如表5.20所示。用这些数据类型存储网络地址比用纯文本类型好，因为这些类型提供输入错误检查和好几种特殊的操作与功能。

表5.20　网络地址类型

名　字	存储空间	描　　述
cidr	12或24字节	IPv4或IPv6网络
inet	12或24字节	IPv4或IPv6网络和主机
macaddr	6字节	MAC地址
macaddr8	8 bytes	MAC addresses (EUI-64 format)

在对inet或cidr数据类型进行排序的时候，IPv4地址总是排在IPv6地址前面，包括那些封装或者映射在IPv6地址里的IPv4地址，比如::10.2.3.4或::ffff::10.4.3.2。macaddr类型存储MAC地址，也就是以太网卡硬件地址。Macaddr8存储EUI-64格式的MAC地址。

（10）位串类型

位串就是一串1和0的字符串。它们可以用于存储和直观化位掩码。PostgreSQL中有两种SQL位类型：bit(n)和bit varying(n)，这里的n是一个正整数。

bit类型的数据必须准确匹配长度n，试图存储短些或者长一些的数据都是错误的。bit varying类型数据是长度最长为n的变长类型；更长的串会被拒绝。写一个没有长度的bit等效于bit(1)，没有长度的bit varying意思是没有长度限制。

（11）文本检索类型

PostgreSQL提供两种支持全文检索的数据类型，tsvector和tsquery。

```
SELECT title
FROM pgweb
WHERE to_tsvector('english', body) @@ to_tsquery('english', 'friend');
```

上述语句代表，在表字段body中查找包含friend字符的记录行，输出记录对应的title字段。

（12）UUID类型

UUID类型用来存储标准的UUID字符。

（13）XML类型

存储XML格式的内容

（14）JSON类型

存储JSON格式的内容。表5.21是JSON对象中数据类型与PostgreSQL数据库中数据类型的对应。

表5.21　数据类型对应

JSON数据类型	PostgreSQL类型	备　　注
string	text	不可以输入Unicode转义值\u0000
number	numeric	不可以输入NaN和infinity
boolean	boolean	只接收小写的true或false
null	（none）	NULL

可以根据具体内容进行筛选，例如。

```
SELECT '{"bar": "baz", "balance": 7.77, "active":false}'::json;
```

（15）数组类型

PostgreSQL允许将字段定义成定长或变长的一维或多维数组。数组类型可以是任何基本类型或用户定义类型。不支持复合类型和域的数组。

首先创建一个由基本类型数组构成的表：

```
CREATE TABLE sal_emp (
    name text,
    pay_by_quarter integer[],
    schedule text[][]
);
```

如上所示，一个数组类型是通过在数组元素类型名后面附加方括号（[]）来命名的。上面的命令将创建一个名叫sal_emp的表，表示雇员名字的name字段是一个text类型字符串，表示雇员季度薪水的pay_by_quarter字段是一个一维integer数组，表示雇员周计划的schedule字段是一个二维text数组。

接着导入数据：

```
INSERT INTO sal_emp
    VALUES ('Bill',
    '{10000, 10000, 10000, 10000}',
    '{{"meeting", "lunch"}, {"training", "presentation"}}');
```

现在我们可以在这个表上进行一些查询。我们演示如何一次访问数组的一个元素,即检索在第二季度薪水发生变化的雇员名:

```
SELECT name FROM sal_emp WHERE pay_by_quarter[1] <> pay_by_quarter[2];
```

数组的下标数字是写在方括弧内的。PostgreSQL默认使用以1为基的数组,也就是说,一个n元素的数组从array[1]开始,到array[n]结束。

(16) 复合类型

复合类型描述一行或者一条记录的结构。它实际上只是一个字段名和它们的数据类型的列表。PostgreSQL允许像简单数据类型那样使用复合类型。比如,一个表的某个字段可以声明为一个复合类型。

下面是两个定义复合类型的简单例子:

```
CREATE TYPE complex AS (
    r double precision,
    i double precision
);
```

语法类似于CREATE TABLE,只是这里只可以声明字段名字和类型;目前不能声明约束(比如NOT NULL)。注意AS关键字是很重要的,没有它,系统会认为这是CREATE TYPE命令,因此会看到奇怪的语法错误。

定义了类型,我们就可以用它创建表:

```
CREATE TABLE on_hand (
    item inventory_item,
    count integer
);
INSERT INTO on_hand VALUES (ROW('fuzzy dice', 42, 1.99), 1000);
```

(17) 区间类型

PostgreSQL包括以下几种区间类型:

- int4range、int8range:整数型离散区间,区间前闭后开。
- numrange:数值连续区间,可以用于描述小数、浮点数或者双精度数值的区间。
- tsrange、tstrange:时间戳(日期加时间)类型的连续区间,秒值部分支持小数。tsrange 不带时区信息,tstzrange 带时区信息。
- datcrange:不带时区信息的日期离散区间。

```
CREATE TABLE reservation (
    room int,
    during tsrange
);
INSERT INTO reservation VALUES (1108, '[2010-01-01 14:30, 2010-01-01 15:30)');
```

(18)域类型

域是基于另一基础类型的用户定义的数据类型,是可选的。它具有将其有效值限制为基础类型所允许的子集的约束。否则,它的行为类似于基础类型。例如,我们可以在整数上创建一个仅接收正整数的域:

```
CREATE DOMAIN posint AS integer CHECK (VALUE > 0);
CREATE TABLE mytable (id posint);
INSERT INTO mytable VALUES(1); -- works
INSERT INTO mytable VALUES(-1); -- fails
```

当基础类型的运算符或函数应用于域值时,域将自动向下转换为基础类型。例如,mytable.id-1的结果被认为是整数类型而不是posint类型。我们可以编写(mytable.id-):: posint将结果转换回posint,从而重新检查域的约束。

(19)对象标识类型

PostgreSQL在内部使用对象标识符(oid)作为各种系统表的主键,但系统不会给用户创建的表增加一个oid系统字段,除非在建表时声明了WITH OIDS或者配置参数default_with_oids。oid类型代表一个对象标识符,它还有几个别名:regproc、regprocedure、regoper、regoperator、regclass、regtype,如表5.22所示。

表5.22 对象标识符类型

名字	引用	描述	数值例子
oid	任意	数字化的对象标识符	564182
regproc	pg_proc	函数名字	sum
regprocedure	pg_proc	带参数类型的函数	sum(int4)
regoper	pg_operator	操作符名	+
regoperator	pg_operator	带参数类型的操作符	*(integer,integer)或-(NONE,integer)
regclass	pg_class	关系名	pg_type
regtype	pg_type	数据类型名	integer

目前oid类型用一个4字节的无符号整数实现。

oid别名类型除了输入和输出过程之外没有自己的操作。这些过程可以为系统对象接收和显示符号名,而不仅仅是oid类型将要使用的行数值。别名类型允许我们简化为对象查找oid值的过程。比如,检查与mytable表相关的pg_attribute行:

```
SELECT * FROM pg_attribute WHERE attrelid = 'mytable'::regclass;
```

而不是下面这样。

```
SELECT * FROM pg_attribute
  WHERE attrelid = (SELECT oid FROM pg_class WHERE relname = 'mytable');
```

(20)pg_lsn类型

pg_lsn类型可用于存储LSN(日志序列号)数据,该数据是指向WAL中某个位置的指针。此类型表示XLogRecPtr和PostgreSQL的内部系统类型。

（21）伪类型

PostgreSQL类型系统包含许多特殊用途的条目，这些条目统称为伪类型。伪类型不能用作列数据类型，但可以用来声明函数的参数或结果类型。如果一个函数不只是简单地接收并返回某种SQL数据类型，就可以使用伪类型。表5.23列出了现有的伪类型。

表5.23　伪类型

名　字	描　述
any	表示一个函数接收任何输入数据类型
anyarray	表示一个函数接收任意数组数据类型
anyelement	表示一个函数接收任何数据类型
anynonarray	表示一个函数接收任意非数组数据类型
anyenum	表示一个函数接收任何枚举数据类型
anyrange	表示一个函数接收任何区间数据类型
anymultirange	表示一个函数接收任何多区间数据类型
anycompatible	表示一个函数接收任何数据类型
anycompatiblearray	表示一个函数接收任何数组数据类型
anycompatiblenonarray	表示一个函数接收任何非数组数据类型
anycompatiblerange	表示一个函数接收任何区间数据类型
anycompatiblemultirange	表示一个函数接收任何多区间数据类型
cstring	表示一个函数接收或者返回一个空结尾的C字符串
internal	表示一个函数接收或者返回一种服务器内部的数据类型
language_handler	一个过程语言调用处理器声明为返回language_handler
fdw_handler	外部数据处理器
record	标识一个函数返回一个未声明的行类型
trigger	一个触发器函数声明为返回 trigger
event_trigger	事件触发器
void	表示一个函数不返回数值
pg_ddl_command	标识数据库操作命令
unknow	未知类型

5.1.3　框架搭建

本章项目选取的是用户权限管理系统，以下为该项目的技术框架部分：

- 开发框架：Spring Boot。
- 数据库：PostgreSQL。
- 后台框架：Spring、Spring MVC、MyBatis、Shrio。
- 前端框架：Thymeleaf、LayUI。

本章项目分为四个大模块，对于框架搭建部分，主要集中于pmrms-framework模块中，关于Spring Boot的配置在pmrms-admin模块中。

Spring、Spring MVC以及MyBatis的整合在第3章中已经做了介绍，在本章中继续沿用。在页面部分，Spring Boot与页面模板解析技术Thymeleaf和LayUI的整合在前面的章节中已提到，本章不再赘述。

接下来讲解在Spring Boot中集成Shrio和PostgreSQL。

```xml
<!--Shrio框架相关依赖-->
<dependency>
    <groupId>org.apache.shiro</groupId>
    <artifactId>shiro-spring</artifactId>
    <version>${shiro.version}</version>
</dependency>
<dependency>
    <groupId>org.apache.shiro</groupId>
    <artifactId>shiro-ehcache</artifactId>
    <version>${shiro.version}</version>
</dependency>
<dependency>
    <groupId>com.github.theborakompanioni</groupId>
    <artifactId>thymeleaf-extras-shiro</artifactId>
    <version>2.0.0</version>
</dependency>
```

引入Shrio依赖后，在Spring Boot配置文件中配置参数，代码如下：

```
# Shiro
shiro:
  operator:
    # 登录地址
    loginUrl: /login
    # 权限认证失败地址
    unauthorizedUrl: /unauth
    # 首页地址
    indexUrl: /index
    # 验证码开关
    captchaEnabled: true
    # 验证码类型，math 数组计算 char 字符
    captchaType: math
  cookie:
    # 设置Cookie的域名，默认为空，即当前访问的域名
    domain:
    # 设置cookie的有效访问路径
    path: /
    # 设置HttpOnly属性
    httpOnly: true
    # 设置Cookie的过期时间，单位为天
    maxAge: 30
    # 设置密钥，务必保持唯一性（生成方式，直接复制到main运行即可）KeyGenerator keygen = KeyGenerator.getInstance("AES"); SecretKey deskey = keygen.generateKey(); System.out.println(Base64.encodeToString(deskey.getEncoded()));
    cipherKey: zSyK5Kp6PZAAjlT+eeNMlg==
  session:
```

```yaml
# session超时时间，-1代表永不过期（默认30分钟）
expireTime: 120
# 同步session到数据库的周期（默认1分钟）
dbSyncPeriod: 1
# 相隔多久检查一次session的有效性，默认就是10分钟
validationInterval: 10
# 同一个用户的最大会话数，比如2的意思是同一个账号最多允许同时两个人登录（默认-1不限制）
maxSession: -1
# 踢出之前登录的/之后登录的用户，默认踢出之前登录的用户
kickoutAfter: false
```

接下来整合PostgreSQL数据库。首先导入依赖：

```xml
<!--连接postgresql数据库驱动包-->
<dependency>
    <groupId>org.postgresql</groupId>
    <artifactId>postgresql</artifactId>
    <version>42.2.5</version>
</dependency>
<!--阿里数据库连接池 -->
<dependency>
    <groupId>com.alibaba</groupId>
    <artifactId>druid-spring-boot-starter</artifactId>
    <version>${druid.version}</version>
</dependency>
```

然后在Spring Boot配置文件中配置数据源：

```yaml
#连接数据库配置信息
spring:
  datasource:
    type: com.alibaba.druid.pool.DruidDataSource
    driverClassName: org.postgresql.Driver
    druid:
      url: jdbc:postgresql://xka:5432/operator
      username: operator
      password: 111111
      # 初始连接数
      initialSize: 5
      # 最小连接池数量
      minIdle: 25
      # 最大连接池数量
      maxActive: 60
      # 配置获取连接等待超时的时间
      maxWait: 6000000
      # 配置间隔多久才进行一次检测，检测需要关闭的空闲连接，单位是毫秒
      timeBetweenEvictionRunsMillis: 70000
      # 配置一个连接在池中最小生存的时间，单位是毫秒
      minEvictableIdleTimeMillis: 30000
      # 配置一个连接在池中最大生存的时间，单位是毫秒
      maxEvictableIdleTimeMillis: 90000
      # 配置检测连接是否有效
```

```yaml
        validationQuery: SELECT 'other'
        #建议配置为true,不影响性能,并且保证安全性。申请连接的时候检测,如果空闲时间大于
timeBetweenEvictionRunsMillis,执行validationQuery检测连接是否有效
        testWhileIdle: true
        #申请连接时执行validationQuery检测连接是否有效,有了这个配置会降低性能
        testOnBorrow: false
        #归还连接时执行validationQuery检测连接是否有效,有了这个配置会降低性能
        testOnReturn: false
        webStatFilter:
          enabled: true
        statViewServlet:
          enabled: true
          # 设置白名单,不填则允许所有访问
          allow:
          url-pattern: /monitor/druid/*
        filter:
          stat:
            enabled: true
            # 慢SQL记录
            log-slow-sql: true
            slow-sql-millis: 1000
            merge-sql: true
          wall:
            config:
              multi-statement-allow: true
```

以上配置完成后,基本框架就搭建完成了。

5.2 项目前期准备

5.2.1 项目需求说明

用户权限管理系统主要实现用户的登录验证及权限控制,包含的功能有操作账号管理、菜单管理、按键管理、组织管理、角色管理和类型管理等。

5.2.2 系统功能设计

用户权限系统的功能结构图如图5.23所示。

- 操作账号管理:对用户登录账号进行管理,能够对账号进行增加、修改、删除操作,设置账号基本信息、账号状态(启用或禁用)、所属组织、所属角色,等等。
- 菜单管理:对系统内的菜单进行管理,对菜单进行增加、修改和删除,并且能够对账号、角色和组织进行权限分配。
- 按键管理:对系统内的按钮权限进行管理,能够实现按钮权限的增加、修改和删除。
- 组织管理:对用户账号所属的组织进行管理,能够对组织进行增加、修改和删除操作。

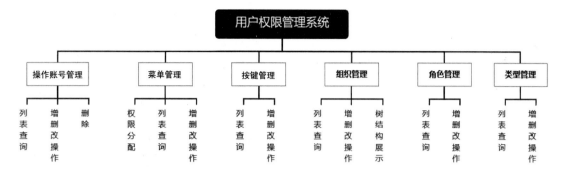

图5.23 用户权限管理系统功能图

- 角色管理：对当前系统内的角色进行管理，能够根据角色类型和所属组织对角色进行增加、修改和删除操作。
- 类型管理：包括组织类型管理和角色类型管理。组织类型用于区分不同类组织，例如公益组织、企业、事业单位或国家机关等；角色类型用于区分不同类角色，如系统角色、业务角色、运维角色、技术角色等。

5.2.3 系统数据库设计

用户权限管理系统主要涉及以下8个表：操作账号表（ta_operator）、角色表（ta_role）、账号角色关联表（ta_person_role）、账号组织关联表（ta_person_rsps）、组织表（ta_organization）、类型表（ta_enttype）、菜单表（ta_menu）、按键表（ta_menu_key）。下面我们逐一介绍这8个表。

- 操作账号表（ta_operator）：主要用于用户的登录、注销以及账号的添加、修改、禁用和删除。
- 角色表（ta_role）：主要用于用户权限控制。
- 账号角色关联表（ta_person_role）：主要用于用户鉴权。
- 账号组织关联表（ta_person_rsps）：主要用于维护账号和组织的关联关系。
- 组织表（ta_organization）：主要用于维护组织相关信息。
- 类型表（ta_enttype）：主要用于维护组织的类型和角色的类型。
- 菜单表（ta_menu）：主要用于维护菜单信息，并用于向账号、组织或角色授权。
- 按键表（ta_menu_key）：主要用于维护菜单下的按钮信息。

由于篇幅关系，这里列出几个关键表的建表语句，其他的建表语句可以通过PostgreSQL的客户端查看。

```
-- Table: common.ta_operator
-- DROP TABLE IF EXISTS common.ta_operator;
--schema:common
--操作账号表
CREATE TABLE IF NOT EXISTS common.ta_operator
(
    operator_code character varying(32) COLLATE pg_catalog."default" NOT NULL,
    operator_name character varying(32) COLLATE pg_catalog."default",
    operator_pwd character varying(32) COLLATE pg_catalog."default",
```

```sql
    operator_sex character varying(1) COLLATE pg_catalog."default",
    operator_mobile character varying(32) COLLATE pg_catalog."default",
    operator_remark character varying(32) COLLATE pg_catalog."default",
    operator_upper character varying(32) COLLATE pg_catalog."default",
    operator_level character varying(32) COLLATE pg_catalog."default",
    operator_rgn_type character varying(32) COLLATE pg_catalog."default",
    operator_state character varying(1) COLLATE pg_catalog."default",
    operator_creattime timestamp without time zone,
    operator_company character varying(32) COLLATE pg_catalog."default",
    CONSTRAINT ha_operator_operator_code_key UNIQUE (operator_code)
)
TABLESPACE pg_default;
ALTER TABLE IF EXISTS common.ta_operator OWNER to postgres;
COMMENT ON COLUMN common.ta_operator.operator_code IS '账号';
COMMENT ON COLUMN common.ta_operator.operator_name IS '姓名';
COMMENT ON COLUMN common.ta_operator.operator_pwd IS '密码';
COMMENT ON COLUMN common.ta_operator.operator_sex IS '性别,M男,F女';
COMMENT ON COLUMN common.ta_operator.operator_mobile IS '手机号';
COMMENT ON COLUMN common.ta_operator.operator_remark IS '备注';
COMMENT ON COLUMN common.ta_operator.operator_upper IS '上级管理账号';
COMMENT ON COLUMN common.ta_operator.operator_level IS '级别';
COMMENT ON COLUMN common.ta_operator.operator_state IS '状态';
COMMENT ON COLUMN common.ta_operator.operator_creattime IS '创建时间';
--角色表
CREATE TABLE IF NOT EXISTS common.ta_role
(
    roleid character varying(63) COLLATE pg_catalog."default" NOT NULL,
    rolename character varying(63) COLLATE pg_catalog."default",
    roleicon character varying(20) COLLATE pg_catalog."default",
    description character varying(255) COLLATE pg_catalog."default",
    typeid character varying(63) COLLATE pg_catalog."default",
    orgid character varying(63) COLLATE pg_catalog."default",
    rolelevel integer,
    CONSTRAINT pk_role PRIMARY KEY (roleid)
)
TABLESPACE pg_default;
ALTER TABLE IF EXISTS common.ta_role OWNER to postgres;
COMMENT ON COLUMN common.ta_role.roleid IS '角色ID';
COMMENT ON COLUMN common.ta_role.rolename IS '角色名称';
COMMENT ON COLUMN common.ta_role.roleicon IS '角色图标';
COMMENT ON COLUMN common.ta_role.description IS '描述';
COMMENT ON COLUMN common.ta_role.typeid IS '类型ID';
COMMENT ON COLUMN common.ta_role.orgid IS '组织ID';
COMMENT ON COLUMN common.ta_role.rolelevel IS '角色级别';
--菜单表
CREATE TABLE IF NOT EXISTS common.ta_menu
(
    id integer NOT NULL DEFAULT nextval('common.ta_menu_seq'::regclass),
    no character varying(64) COLLATE pg_catalog."default" NOT NULL,
    name character varying(30) COLLATE pg_catalog."default",
    perms character varying(512) COLLATE pg_catalog."default",
    url character varying COLLATE pg_catalog."default",
    parent_no character varying COLLATE pg_catalog."default",
```

```
        nodes integer,
        data_id character varying(20) COLLATE pg_catalog."default",
        icon character varying COLLATE pg_catalog."default",
        menu_type character varying(20) COLLATE pg_catalog."default",
        hidden character varying(50) COLLATE pg_catalog."default",
        pos double precision,
        createby character varying(30) COLLATE pg_catalog."default",
        modifyby character varying(30) COLLATE pg_catalog."default",
        createtime timestamp(0) without time zone,
        modifytime timestamp(0) without time zone
)
TABLESPACE pg_default;
ALTER TABLE IF EXISTS common.ta_menu  OWNER to postgres;
COMMENT ON COLUMN common.ta_menu.id IS '菜单id';
COMMENT ON COLUMN common.ta_menu.no IS '编号';
COMMENT ON COLUMN common.ta_menu.name IS '名称';
COMMENT ON COLUMN common.ta_menu.perms IS '权限';
COMMENT ON COLUMN common.ta_menu.url IS '操作url';
COMMENT ON COLUMN common.ta_menu.parent_no  IS '父级编号';
COMMENT ON COLUMN common.ta_menu.icon IS '图标';
COMMENT ON COLUMN common.ta_menu.menu_type IS '类型';
COMMENT ON COLUMN common.ta_menu.createby IS '创建者';
COMMENT ON COLUMN common.ta_menu.modifyby IS '修改者';
COMMENT ON COLUMN common.ta_menu.createtime IS '创建时间';
COMMENT ON COLUMN common.ta_menu.modifytime IS '修改时间';
```

5.2.4 系统文件说明

本项目的总体工程系统文件说明如图5.24所示。

整个系统使用Maven管理，分为四个模块，分别是pmrms-admin、pmrms-common、pmrms-framework、pmrms-system。下面对各个部分作简要说明。

（1）pmrms-admin

这是系统的Web模块，即项目启动模块，该模块说明如图5.25所示。

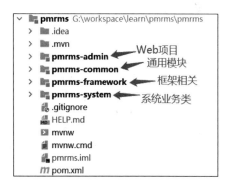

图5.24　总体工程说明

- PmrmsAdminApplication：Spring Boot启动类。
- controller：控制器类，处理接口。
- web：Web相关的控制器类，即首页跳转、登录、注销等。
- 前端相关：所有的静态资源（static）和页面信息（templates）。

（2）pmrms-common

这是系统的公用基础模块，包括一些基础配置、常量配置、通用工具、通用注解、全局异常和页面配置等，如图5.26所示。

（3）pmrms-framework

这是系统框架相关的公用部分，包括Spring框架的基础配置、切面信息、数据源设置、拦截器、Shrio框架配置等，如图5.27所示。

图5.25　admin模块说明

图5.26　common模块说明

（4）pmrms-system

这是系统业务相关的模块，包括实体类、数据接口以及业务实现类，如图5.28所示。

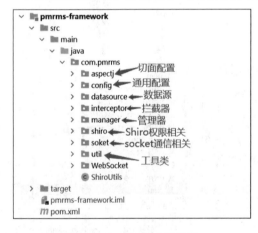

图5.27　framework模块说明

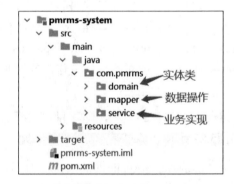

图5.28　system模块说明

5.3　项目前端设计

5.3.1　登录

登录时，输入用户名、密码和验证码，校验正确即可跳转至管理控制台，否则返回登录页，并给出失败提示，如图5.29～图5.31所示。

图5.29　登录页面　　　　图5.30　密码错误　　　　图5.31　验证码错误

验证代码如下：

```
<div class="form-group">
    <div class="col-lg-12">
        <input class="layui-input" id="checkCodeYZM" name="checkCode" placeholder="验证码"
            lay-verify="myRequired" type="text" autocomplete="off">
        <canvas id="canvas" class="canvasCode">
        </canvas>
    </div>
</div>
```

这里的验证使用的是LayUI支持的自定义验证，组件的属性为lay-verify，该部分内容在第2章中有详细介绍。myRequired是自定义的验证规则，这里比较简单，只校验了必填，我们还可以自定义文本规则，例如邮箱、IP地址等。

```
form.verify({
    myRequired: function (value, item) { //value, 表单的值；item, 表单的DOM对象
        if (!value) {
            return '必填项不能为空';
        }
    }
});
```

这里重点讲一下验证码的生成与校验。

在登录页首次加载进入时，系统自动生成一次验证码，如果用户单击验证码图片，则重新生成验证码。

```
drawshow_num(show_num);//初始化加载时自动生成一次
/**单击验证码重新生成*/
$("#canvas").on('click', function () {
    drawshow_num(show_num);
});
```

本章验证码选用的是纯数字。验证码中真正有效的信息是数字，但为了防止刷屏或者恶意攻击，我们会对验证码做一些模糊处理，使它看起来不那么容易被识别，目的是对机器进行干扰。要模糊这个验证码，就要用到图片。所以这里使用了canvas组件来显示验证码。

验证核心是draw()方法，该方法传入两个参数，一个存放验证码按位分割后的数组，一个存放验证码数字。

```javascript
//向接口请求验证码数字
function drawshow_num(show_num) {
    $.ajax({
        cache: true,
        type: "GET",
        url: "/randomNum",
        async: false,
        success: function (k) {
            draw(show_num, k);//画验证码
        }
    });
}
function randomColor() {//得到随机的颜色值==>把颜色设置为黑色，看得清楚
    var r = Math.floor(0);
    var g = Math.floor(0);
    var b = Math.floor(0);
    return "rgb(" + r + "," + g + "," + b + ")";
}
//画验证码
function draw(show_num, sCode) {
    var canvas_width = 100;
    var canvas_height = 36;
    var canvas = document.getElementById("canvas");//获取canvas的对象，类比演员
    var context = canvas.getContext("2d");//获取canvas画图的环境，类比演员表演的舞台
    canvas.width = canvas_width;
    canvas.height = canvas_height;
    var aCode = sCode.split(",");
    //画验证码数字
    for (var i = 0; i <= 3; i++) {
        var deg = Math.random() * 30 * Math.PI / 180;//产生0~30的随机弧度
        var txt = aCode[i];//得到随机的一个内容
        show_num[i] = txt.toLowerCase();
        var x = 10 + i * 20;//文字在canvas上的x坐标
        var y = 20 + Math.random() * 8;//文字在canvas上的y坐标
        context.font = "bold 26px 楷体";
        context.translate(x, y);
        context.rotate(deg);
        context.fillStyle = randomColor();
        context.fillText(txt, 0, 0);
        context.rotate(-deg);
        context.translate(-x, -y);
    }
    for (var i = 0; i <= 5; i++) { //验证码上显示线条
        context.strokeStyle = randomColor();
        context.beginPath();
        context.moveTo(Math.random() * canvas_width, Math.random() * canvas_height);
```

```
            context.lineTo(Math.random() * canvas_width, Math.random() *
canvas_height);
            context.stroke();
        }
        for (var i = 0; i <= 30; i++) { //验证码上显示小点
            context.strokeStyle = randomColor();
            context.beginPath();
            var x = Math.random() * canvas_width;
            var y = Math.random() * canvas_height;
            context.moveTo(x, y);
            context.lineTo(x + 1, y + 1);
            context.stroke();
        }
    }
```

5.3.2 控制台首页

登录成功后，进入控制台首页。这里需要展示的信息有当前登录状态、导航栏、菜单栏，如图5.32所示。

图5.32 控制台首页

代码如下：

```
<div class="sl-sideleft-menu">
  <ul id="nav">
    <li>
       <a href="javascript:;" class="sl-menu-link active menuItem" data-id="0" data-url="/openMainHtml">
          <div class="sl-menu-item">
             <i class="menu-item-icon icon ion-ios-home-outline tx-18"></i>
             <span class="menu-item-label">首页</span>
          </div>
       </a>
    </li>
    //遍历显示菜单信息
       <li th:each="m,stat:${session.menu}" th:class="${m.menuType}">
```

```html
                <a href="javascript:;" th:data-nodes="${m.nodes}" th:data-title="${m.name}" th:data-id="${m.dataId}" th:data-url="${m.url}" class="sl-menu-link menuItem" >
                    <div class="sl-menu-item">
        //菜单图标
                        <i th:class="${m.icon}"></i>
                        <span class="menu-item-label" th:text="${m.name}"></span>
        //是否存在子菜单
                        <i th:if="${ m.listMenu.size() > 0 or m.nodes > 0 }" class="menu-item-arrow fa fa-angle-down"></i>
                    </div><!-- menu-item -->
                </a>
        //子菜单
                <ul th:if="${m.listMenu.size() > 0}" class="sl-menu-sub nav flex-column">
                    <li class="nav-item" th:each="m2,stat:${m.listMenu}" th:class="${m2.menuType}">
                        <a href="javascript:;" class="sl-menu-link nav-link menuItem" th:data-nodes="${m2.nodes}"
                           th:data-title="${m2.name}" th:data-id="${m2.dataId}" th:data-url="${m2.url}">
                            <span th:text="${m2.name}"></span>
                            <i th:if="${m2.listMenu.size() > 0}" class="menu-item-arrow fa fa-angle-down" style="margin-left: 10px;"></i>
                        </a>
                        <ul th:if="${m2.listMenu.size() > 0}" class="sl-menu-sub nav flex-column">
                            <li class="nav-item" th:each="m3,stat:${m2.listMenu}">
                                <a href="javascript:;" class="nav-link menuItem" th:data-title="${m3.name}" th:url="${m3.url}" th:text="${m3.name}"></a>
                            </li>
                        </ul>
                    </li>
                </ul>
            </div>
```

菜单数据由接口返回，包含菜单名称、菜单类型、图标、URL等。

这个页面会连接WebSocket，用于及时获取会话登录状态，若会话失效，则强制退出登录，返回登录页。这是一个全局操作。

```javascript
var socketUrl=localhostPaht+"/imserver/"+getSessionId();
socketUrl=socketUrl.replace("https","ws").replace("http","ws");
if(socket!=null){
    socket.close();
    socket=null;
}
socket = new WebSocket(socketUrl);
//打开事件
socket.onopen = function() {
```

```
        console.log("websocket已打开");
        //socket.send("这是来自客户端的消息" + location.href + new Date());
    };
    //获得消息事件
    socket.onmessage = function(msg) {
        console.log(msg.data);
        admin_del(msg.data);
        //发现消息进入，开始处理前端触发逻辑
    };
    //关闭事件
    socket.onclose = function() {
        console.log("websocket已关闭");
    };
    //发生了错误事件
    socket.onerror = function() {
        console.log("websocket发生了错误");
    }
```

5.3.3 操作账号管理

操作员即系统用户，用于登录和认证的账号。操作账号管理包含操作员列表查询，新增、修改、删除操作员账号等功能。操作员列表包含账号的基本信息以及所属的角色和公司，如图5.33所示。

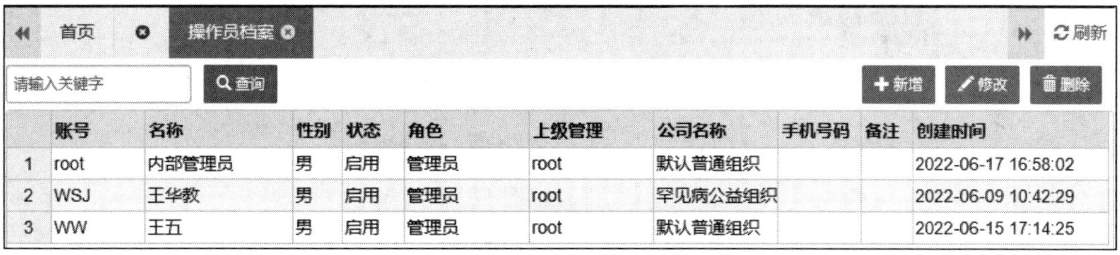

图5.33 操作员档案

新增操作员页面包含账号基本信息，包括账号、名称、性别、手机号码、密码、账户状态、所属组织和角色等，如图5.34所示。其中账号、名称和密码为必填项，编辑页面与添加页面类似，区别在于编辑时不能够编辑密码。

图5.34 新增操作员

5.3.4 菜单管理

菜单管理用于管理导航菜单。菜单管理包含列表查询、新增、修改和删除菜单，权限分配，等等，如图5.35所示。

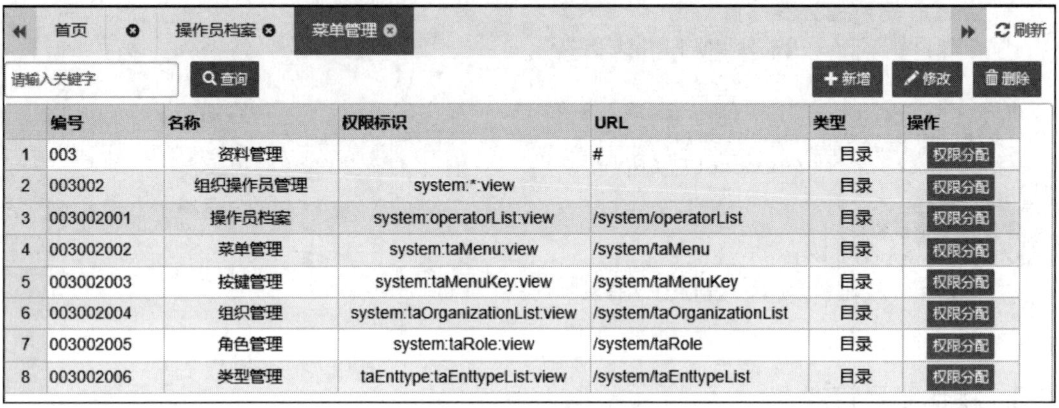

图5.35 菜单管理

新增菜单时，页面如图5.36所示。该页面包含菜单基本信息，包括菜单类别、所属角色、菜单名称、URL、图标、父级菜单、排序、显示或隐藏等。修改时，其页面与添加类似。

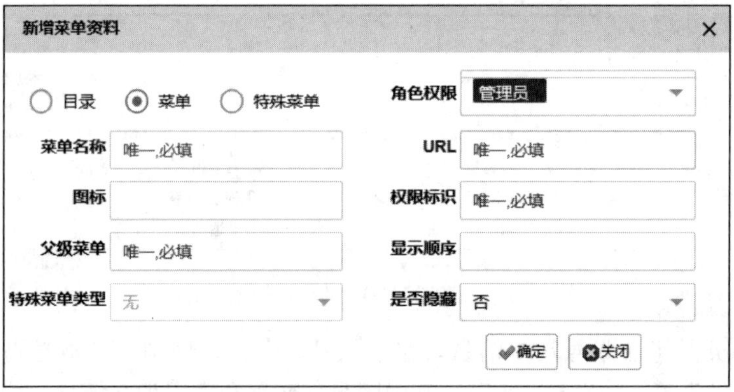

图5.36 新增菜单

分配权限时，弹出权限设置页面，为对应的用户、角色或组织分配权限，使得这些对象拥有这个菜单资源的访问权限，如图5.37所示。

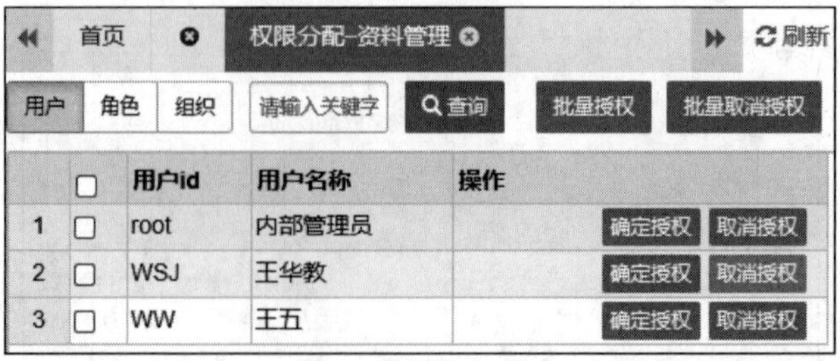

图5.37 权限分配

5.3.5 按键管理

按键管理是对页面上的按钮进行管理。在这里配置的按钮，如果操作者没有对应的权限，是无法进行操作的。按键管理包含按键列表，新增、修改和删除按键项目，如图5.38所示。

图5.38 按键管理

新增按键时，页面如图5.39所示。该页面包含的基本信息有按键名称、按键URL、所属菜单以及权限标识。权限标识用来表明该按键属于哪类操作，view代表查看，add代表新增，remove代表删除，edit代表编辑。

图5.39 新增按键

5.3.6 组织管理

组织表示某一社会组织或团体，也可以是公司或企业，或者任何单位。它可以是单独的组织，也可以是一个拥有上下级关系的组织结构。它表示操作者实际所属组织，便于进行权限控制，组织管理页面如图5.40所示。

图5.40 组织管理页面

新增组织结构时，页面如图5.41所示。该页面包含以下内容：组织名称、上级组织、类型、描述以及组织级别。修改时，其页面与新增类似。

图5.41 新增组织结构

组织类型支持下拉框选择，该数据可以在类型管理中进行管理和维护。

5.3.7 角色管理

角色代表一系列权限的合集。角色与操作者相关联，被赋予某一角色的操作者拥有该角色下配置的所有权限。角色管理包含列表查询、新增、修改和删除角色，如图5.42所示。

图5.42　角色管理

在列表查询页面左侧是角色类型树，直观展示当前系统中所有角色类型以及对应的层级关系。该类型可在类型管理中进行管理与维护。

新增角色时，页面如图5.43所示。该页面包含角色相关的基本信息，包括角色名称、类型、所属组织、角色描述以及角色级别。

选择所属组织时，会弹出选择树页面，如图5.44所示。

图5.43　新增角色

图5.44　选择上级组织

5.3.8 类型管理

类型是指角色类型和组织类型。在这里可以动态维护对应的类型，所做操作引起的数据改变在其他引用到该数据的地方都会进行动态刷新。类型管理包含列表查询，新增、修改、删除类型，如图5.45所示。

在列表页，可以在上方的选择下拉框中筛选出对应的类型，如图5.46所示。

图5.45　组织类型管理

图5.46　类型筛选

新增类型时，页面如图5.47所示。该页面包含的基本信息有类型名称、描述和标识（即组织或角色类型区分）。

图5.47　新增类型

5.3.9　分页展示

该分页为全局分页展示，如图5.48所示。系统中凡是用到分页的地方均引用该样式和内容。

图5.48　分页

5.4　项目后端实现

5.4.1　登录认证和权限认证

本章项目采用Shrio权限认证框架进行登录权限认证。使用该框架前，我们需要做些相应的配置。

首先要在Spring配置文件application.yml中加入Shrio相关的配置。这部分在5.1.3节框架搭建部分有详细描述。

其次，要编写自定义认证器OperatorRealm，用于认证当前用户信息。该认证器需要验证用户信息以及用户权限信息，所以要提前编写用户、角色和权限相关的Service方法。在认证器中，有两个重要的方法，一个是登录认证，一个权限认证。登录认证用来校验账号和密码是否正确，权限认证用来校验用户是否拥有当前操作的权限。

登录认证的流程：从AuthenticationToken对象中解析出账号和密码，调用Service中的login()方法与数据库中的账号和密码作比对，若认证成功则返回用户对象信息，若不成功，则根据不同的失败情况抛出对应的异常，便于上层调用者进行捕捉。

```
/**
 * 登录认证
 * @param token
 * @return
 * @throws AuthenticationException
 */
@Override
protected AuthenticationInfo doGetAuthenticationInfo(AuthenticationToken token)
throws AuthenticationException {
```

```java
        UsernamePasswordToken upToken = (UsernamePasswordToken) token;//获取当前用户信息token
        String uid = upToken.getUsername();              //获取账号
        String pwd = "";
        if (upToken.getPassword() != null)
        {
            pwd = new String(upToken.getPassword());      //获取密码
        }
        TaOperator TaOperator = null;
        try {
            TaOperator = operatorService.login(uid,pwd);//进行登录操作,与数据库中的用户信息作比对
        }catch (TaOperatorNullException e){              //账号不存在
            throw new UnknownAccountException(e.getMessage(),e);
        }catch (TaOperatorStateISException e){           //用户被禁用
            throw new UnknownAccountException(e.getMessage(),e);
        }catch (TaOperatorNotPasswordException e){       //密码不正确
            throw new UnknownAccountException(e.getMessage(),e);
        }catch (Exception e) {                           //其他异常信息
            e.printStackTrace();
            throw new TaOperatorSystemException();
        }
        //登录认证成功,返回认证信息
        SimpleAuthenticationInfo info = new SimpleAuthenticationInfo(TaOperator, pwd, getName());
        return info;
    }
```

权限认证的一般流程：首先获取当前用户的账号,根据账号调用Service方法查询对应的角色和权限集合。然后将角色信息和权限信息存入SimpleAuthorizationInfo对象并返回。如果遇到需要特殊处理的用户,那么在查询角色和权限前,对账号进行特殊处理,例如下述代码中的root和ha用户。如果遇到这两个用户,那么默认它们拥有全部的权限。

```java
    /**
     * 权限认证
     * @param principals
     * @return
     */
    @Override
    protected AuthorizationInfo doGetAuthorizationInfo(PrincipalCollection principals) {
        String operatorCode = ShiroUtils.getOperatorCode();           //获取用户账号
        SimpleAuthorizationInfo info = new SimpleAuthorizationInfo();
        if("root".equals(operatorCode) || "ha".equals(operatorCode)){//超级管理员或特定的ha账号,默认授予对应的角色以及角色下的所有权限
            info.addRole("root");
            info.addRole("ha");
            info.addStringPermission("*:*:*");
        }else {                              //其他账号,根据账号去查询对应的角色和权限菜单
            List<TaMenu> haMenuList = taMenuService.selectMenuByOperatorCode(operatorCode);
```

```java
            List<TaRole> taRoles = taRoleService.selectTaRoleByCode(operatorCode);
            Optional<TaRole> role = taRoles.stream().filter(m -> m.getRoleId().equals
("R0000000001")).findFirst();
            for (TaRole taRole:taRoles){
                info.addRole(taRole.getRoleId());
            }
            if(role.isPresent()){          //如果是超级管理员的角色,则直接授予全部权限
                info.addStringPermission("*:*:*");
            }else {                        //没有超级管理员角色的,根据角色对应的权限进行授予
                for (TaMenu taMenu : haMenuList) {
                    TaFunRole taFunRole = taMenu.getTaFunRole();
                    if(taFunRole != null) {
                        info.addStringPermission(taFunRole.getPermissions());
                    }
                }
            }
        }
        return info;
    }
```

以上内容完成后,可以根据5.1.1节中提到的内容进行其他个性化的配置,也可以参考本章项目代码。接下来我们重点看ShrioConfig配置类。在这个配置类中,有两个比较重要的Bean配置,一个是会话管理器,另外一个是过滤器工厂Bean。会话管理器中存储着当前用户对应Session的相关信息,在会话结束前或Session失效前,用户信息一直存在管理器中。在会话管理器中使用到的Bean在该配置类中即可注入,它们的功能可以参考5.1.1节中关于Shrio框架的介绍。这里有一个全局Session超时时间,代表当前会话消息的有效期,在开发阶段如果不想频繁登录,可以取消该设置,这样Session就不再失效。切记,在生产环境中不可采取此种做法。

```java
/**
 * 会话管理器
 */
@Bean
public OnlineWebSessionManager sessionManager() {
    OnlineWebSessionManager manager = new OnlineWebSessionManager();
    Collection<SessionListener> listeners = new ArrayList<SessionListener>();
    // 加入缓存管理器
    //manager.setCacheManager(getEhCacheManager());
    //配置监听
    listeners.add(sessionListener());
    manager.setSessionListeners(listeners);
    manager.setSessionIdCookieEnabled(true);
    manager.setSessionIdCookie(sessionIdCookie());
    // 自定义SessionDao
    manager.setSessionDAO(sessionDAO());
    // 自定义sessionFactory
    manager.setSessionFactory(sessionFactory());
    // 设置全局session超时时间
    manager.setGlobalSessionTimeout(expireTime * 60 * 1000);
    manager.setDeleteInvalidSessions(true);
    //是否开启扫描
```

```java
    manager.setSessionValidationSchedulerEnabled(true);
    // 去掉地址栏后面的sessionid
    manager.setSessionIdUrlRewritingEnabled(false);
    //检测扫描信息时间间隔,单位为毫秒
    manager.setSessionValidationInterval(5000);
    return manager;
}
```

过滤器工厂用来配置Shrio认证的规则，包括：如果当前用户登录失败，需要跳转的页面；不需要进行认证拦截的静态资源链接、特殊链接以及登录、注销页面。

```java
/**
 * Shiro过滤器配置
 */
@Bean
public ShiroFilterFactoryBean shiroFilterFactoryBean(SecurityManager securityManager)
{
    ShiroFilterFactoryBean shiroFilterFactoryBean = new ShiroFilterFactoryBean();
    // Shiro的核心安全接口,这个属性是必须的
    shiroFilterFactoryBean.setSecurityManager(securityManager);
    // 身份认证失败,则跳转到登录页面
    shiroFilterFactoryBean.setLoginUrl(loginUrl);
    shiroFilterFactoryBean.setUnauthorizedUrl("/403.html");
    // Shiro连接约束配置,即过滤器链的定义
    LinkedHashMap<String, String> filterChainDefinitionMap = new LinkedHashMap<>();
    // 对静态资源设置匿名访问
    filterChainDefinitionMap.put("/favicon.ico**", "anon");
    filterChainDefinitionMap.put("/common/**", "anon");
    filterChainDefinitionMap.put("/images/**", "anon");
    filterChainDefinitionMap.put("/ajax/**", "anon");
    filterChainDefinitionMap.put("/plugins/**", "anon");
    filterChainDefinitionMap.put("/system/**", "anon");
    // 退出 logout地址,使用shiro清除session
    filterChainDefinitionMap.put("/logout", "logout");
    // 不需要拦截的访问
    filterChainDefinitionMap.put("/login", "anon");
    // 随机数验证码不需要拦截的访问
    filterChainDefinitionMap.put("/randomNum", "anon");
    //微信公众号地址不需拦截
    filterChainDefinitionMap.put("/wx", "anon");
    //外地接口不需要拦截
    filterChainDefinitionMap.put("/interface/readMeter.asp","anon");
    filterChainDefinitionMap.put("/api/v1/**","anon");
    Map<String, Filter> filters = new LinkedHashMap<>();
    //filters.put("prems",shiroPermissionsFilter());
    filters.put("onlineSession", onlineSessionFilter());
    filters.put("syncOnlineSession", syncOnlineSessionFilter());
    //filters.put("urlPermissionsFilter",urlPermissionsFilter());
    filters.put("logout",logoutFilter());
```

```
    shiroFilterFactoryBean.setFilters(filters);
    // 所有请求需要认证
    filterChainDefinitionMap.put("/**", "user,onlineSession,syncOnlineSession");
    shiroFilterFactoryBean.setFilterChainDefinitionMap(filterChainDefinitionMap);
    return shiroFilterFactoryBean;
}
```

编写完配置类之后,我们需要在该配置上加上@Configuration注解,这样Spring才能够自动扫描Shrio的配置,并做初始化加载。至此,Shrio框架中与登录和权限认证相关的配置已经设置完毕。接下来我们看一下如何在Controller代码中进行调用。

首先我们需要构造用于认证的Token对象,这个对象中包含用户名和密码。然后得到当前主体信息Subject,接着进行登录,如果登录成功,则返回"true",如果登录失败,则给出提示信息。

```
/**
 * 登录的方法
 * @param uid
 * @param pwd
 * @param mmap
 * @return
 */
@PostMapping("/login")
@ResponseBody
public String login(@RequestParam( value = "uid") String uid,
                    @RequestParam( value = "pwd") String pwd
                    , ModelMap mmap , HttpSession session){
    //构造登录认证Token对象
    UsernamePasswordToken token = new UsernamePasswordToken(uid, pwd);
    //得到当前的主体对象
    Subject subject = SecurityUtils.getSubject();
    try{
        subject.login(token);          //登录操作,如果成功则进入下一步代码,如果失败则进入catch代码块
        session.setAttribute("publicUid",uid);
        return "true";
    }catch (AuthenticationException e){              //捕捉登录失败异常
        String msg = "";
        if (StringUtils.isNotEmpty(e.getMessage())){    //失败信息
            msg = e.getMessage();
        }
        return msg;
    }
}
```

5.4.2 验证码生成

登录时需要输入验证码,该验证码可由前端页面直接生成,也可以通过后台生成。以下是后台生成4位数验证码的参考方法:

```java
//后台只生成随机数
@GetMapping("/randomNum")
@ResponseBody
public void findRandom (HttpServletResponse response) throws IOException{
    // 验证码字符个数
    int codeCount = 4;
    char[] codeSequence = { '0', '1', '2', '3', '4', '5', '6', '7', '8', '9' };
    // 创建一个随机数生成器类
    Random random = new Random();
    // randomCode用于保存随机产生的验证码,以便用户登录后进行验证
    StringBuffer randomCode = new StringBuffer();
    //获取数组的长度
    int aLength = codeSequence.length;
    for (int i = 0; i < codeCount; i++) {
        // 得到随机产生的验证码数字
        String strRand = String.valueOf(codeSequence[random.nextInt(aLength)]);
        // 将产生的四个随机数组合在一起
        randomCode.append(strRand).append(",");
    }
    PrintWriter out = response.getWriter();
    out.print(randomCode);
}
```

这里用到了Random类。在Java中有两个Random类，java.lang.Math.Random和java.util.Random。这里使用的是java.util.Random。它是一个随机数生成器，可以随机生成各种类型的数据，例如float、double、long、int、boolean等，它还能生成各种数据类型的数值流。

Random类的常用方法如表5.24所示。

表5.24 Random类的常用方法

方法名	返回类型	说　　明
nextBoolean()	boolean	返回一个布尔型随机数
nextDouble()	double	返回一个在[0,1)区间的double数值
nextFloat()	float	返回一个在[0,1)区间的float数值
nextGaussian()	double	返回一个高斯/正态分布double值，产生的数字是符合标准正态分布的
nextInt()	int	返回一个int随机数
nextInt(int bound)	int	返回一个在区间[0,bound)之间的int随机数
nextLong()	long	返回一个long随机数
doubles()	DoubleStream	创建一个无穷大的double类型的数值流，每个值在[0,1)区间
ints()	IntStream	创建一个无穷大的随机int值流。
longs()	LongStream	创建一个无穷大的随机long值流。

5.4.3　操作账号管理

搜索操作员时，首先接收前端传递的参数对象TaOperator，然后传递参数到Service方法中，根据传递的参数构造查询条件查询出对应的列表结果。

```java
@RequestMapping("/operatorListJson")
@ResponseBody
public Object operatorListJson(TaOperator TaOperator) {
    startPage();//设置分页参数
    TaOperator.getParams().put("judgeFrom", SqlCondition.judgeFrom());//根据不同角色拼接SQL
    //调用Service方法查询结果
    List<TaOperator> TaOperatorsList = taOperatorService.selectTaOperatorList(TaOperator);
    //封装结果
    return getDataTable(TaOperatorsList);
}
```

返回数据时使用了getDataTable方法，这是本项目中后台返回分页表格数据时的通用处理方法。在该方法中，传入查询结果（通常是个List）构造出对应结果集的Map，这个Map中包含分页相关信息以及数据结果集。本章项目的所有分页数据封装都使用此方式。

```java
//通用结果集封装方法
protected Map<String,Object> getDataTable(List<?> list)
{
    Map<String,Object> m = new HashMap<String, Object>();
    //
    long records = new PageInfo(list).getTotal();
    //分页参数设置
    PageDomain pageDomain = TableSupport.buildPageRequest();
    try {
        pageDomain.setTotal(Integer.parseInt(Long.toString(records)));
    } catch (NumberFormatException e) {
        logger.error(e.getMessage());
    }
    //封装分页参数
    m.put("page",pageDomain.getPage());
    m.put("rows",list);//数据结果集
    m.put("total",pageDomain.getTotal());//总条数
    return m;
```

新增操作员时，前端传递操作员信息到接口，接口接收参数后首先新增操作员信息，然后再添加操作员和角色的绑定关系，接着添加操作员和组织的绑定关系。这些操作要放在同一事务中进行，防止程序运行中间停止，而导致数据记录不全、产生脏数据。在执行新增操作员的SQL时，除了操作员基本信息外，还需要插入当前时间，这里用到了SQL中的now()函数。

```xml
<insert id="addTaOperator">
    insert into common.ta_operator (
    <if test="operatorCode != null and operatorCode != ''">operator_code,</if>
    <if test="operatorName != null and operatorName != ''">operator_name,</if>
    <if test="operatorPwd != null and operatorPwd != '' ">operator_pwd,</if>
    <if test="operatorSex != null and operatorSex != '' ">operator_sex,</if>
    <if test="operatorMobile != null and operatorMobile != ''">operator_mobile,</if>
    <if test="operatorRemark != null and operatorRemark != ''">operator_remark,</if>
```

```xml
        <if test="operatorUpper != null and operatorUpper != ''">operator_upper,</if>
        <if test="operatorState != null and operatorState != ''">operator_state,</if>
        operator_creattime
        )values(
        <if test="operatorCode != null and operatorCode  != ''">#{operatorCode},</if>
        <if test="operatorName != null and operatorName != ''">#{operatorName},</if>
        <if test="operatorPwd != null and operatorPwd != ''">#{operatorPwd},</if>
        <if test="operatorSex != null and operatorSex != ''">#{operatorSex},</if>
        <if test="operatorMobile != null and operatorMobile != ''">#{operatorMobile},
</if>
        <if test="operatorRemark != null and operatorRemark != ''">#{operatorRemark},
</if>
        <if test="operatorUpper != null and operatorUpper != ''">#{operatorUpper},</if>
        <if test="operatorState != null and operatorState != ''">#{operatorState},</if>
        now()
        )
</insert>
```

5.4.4 菜单管理

菜单管理部分除了基本的列表查询、新增、修改和删除功能外，还有一个特殊的功能叫权限分配。菜单创建好之后，能被哪些用户、角色或组织查看和访问，都在权限分配里进行设置。

权限分配设置的接口逻辑如下：

首先，后端接收前端传递的参数，该参数包含以下内容：

- menuId：当前菜单id。
- objId：菜单所赋予对象id（用户、角色、组织）。
- objType：objId所属类型，1：用户，2：角色，3：组织。
- permissions：具体权限，例如views代表查看。

例如，将资料管理设置为用户WSJ可以查看，发送的参数为：

```
{
  "objId": "WSJ",
  "objType": "1",
  "menuId": 22,
  "permissions": "view"
}
```

后端接收时使用Map方法，整个方法使用@Transactional注解表示支持事务。

```
@Transactional(rollbackFor = Exception.class)
@PostMapping("/add")
@ResponseBody
public String taFunRoleAddPost(@RequestBody Map<String, Object> map) {
    ...
}
```

其次，后端接收到参数后，先进行参数判断，然后查找数据库作对比，不存在关系的直接新建，已存在的关系进行更新。

```
    Object obj = map.get("objId");//接受权限分配的对象id
    if (obj != null && !StringUtils.isEmpty(obj.toString())) {
        String permissions = map.get("permissions") == null ? null :
map.get("permissions").toString();//权限
        Integer id = map.get("menuId") == null ? null : (Integer) map.get("menuId");//
菜单id
        String objType = map.get("objType") == null ? null :
map.get("objType").toString();//对象类型，1用户，2角色，3组织
        //存放权限关系的查询条件
        TaFunRole taFunRole = new TaFunRole();
        TaMenu taMenu = taMenuService.selectMenuById(id);    //判断菜单是否存在
        if (taMenu == null)
            return "菜单不存在 || 菜单权限标识为空";
        if (id != null) {
            taFunRole.setTaMenu(taMenu);                      //查询条件：设置菜单
        }
        taFunRole.setObjId(obj.toString());                   //查询条件：设置对象id
        if (permissions != null) {
            if (taMenu != null) {
                //查询条件：设置权限内容
                taFunRole.setPermissions(taMenu.getPerms() + permissions);
            }
        }
        taFunRole.setObjType(objType);
        //找到对象id和菜单id对应的权限关系
        List<TaFunRole> taFunRoleList = taFunRoleService.selectTaFunRoleAll
(taFunRole);
        if (taFunRoleList.size() > 0) {
            TaFunRole taFunRole1 = taFunRoleList.get(0);        //已存在的权限关联关系
            String permissions2 = taFunRole1.getPermissions();  //已存在的权限关系
            String permissions3 = permissions2.substring(0,
permissions2.lastIndexOf(":") + 1);                              //权限归属模块
            taFunRole1.setPermissions(permissions3 + permissions);//设置新的权限标识
            //新的权限标识与数据库已存在的权限标识作对比，不一样时，进行更改
            if (!taFunRole1.getPermissions().equals(permissions2)) {
                taFunRole1.setModifyBy(ShiroUtils.getOperatorCode());
                taFunRole1.setModifyTime(new Date());
                taFunRoleService.updateTaFunRole(taFunRole1);
            }
        } else {//如果不存在已有的关系，那么直接创建新的关系存入数据库
            taFunRole.setCreateBy(ShiroUtils.getOperatorCode());
            taFunRole.setCreateTime(new Date());
            taFunRoleService.insertTaFunRole(taFunRole);
        }
    }
}
```

5.4.5 组织管理

组织管理主要实现组织对象的查询、新增、修改和删除。

新增组织时，需要传递的参数有：

- orgName：组织名称。
- Description：描述。
- uporgId：上级组织。
- orgLevel：组织级别。
- typeIds：组织类型数组。

接收到参数后，首先查找数据库中最大的组织编号，如果查询结果为空，则说明数据库中没有组织数据，这时编号从1开始；如果存在，则将最大编号加1作为新的编号，再将组织类型数组转换成String类型插入数据库。代码如下：

```
/**
 * 新增组织
 * @param taOrganization 组织基本信息
 * @param typeIds    组织类型数组
 * @return
 */
@RequestMapping("/addTaOrganization")
@ResponseBody
public Object addTaOrganization(TaOrganization taOrganization
        ,@RequestParam(value = "typeIds[]",required = false) String[] typeIds) {
    String r = "O";//用于编号补0
    //找到目前数据库中组织的最大编号
    TaOrganization taOrgIdMax = taOrganizationService.getTaOrgIdMax();
    if (taOrgIdMax == null) {
        //如果不存在，那么从1开始
        String rNumbers = "0000000000";
        String rNumber = String.format("%0" + rNumbers.length() + "d", Integer.parseInt(rNumbers) + 1);
        taOrganization.setOrgId(r + rNumber);
    } else {
        //如果存在，那么最大编号加1，作为新的编号
        String number = taOrgIdMax.getOrgId().substring(1, 11);
        String rNumber = String.format("%0" + number.length() + "d", Integer.parseInt(number) + 1);
        taOrganization.setOrgId(r + rNumber);
    }
    //所属组织类型数组转换
    String typeIdsHandled = Arrays.stream(typeIds).map(i -> i.toString()) //必须将普通数组转换成数据流才能在map里面toString
            .collect(Collectors.joining(","));
    taOrganization.setTypeId(typeIdsHandled);
    taOrganization.setFlag("1");
    taOrganization.setUrl("");
    //执行插入操作
    Integer addTaOrganization = taOrganizationService.insertTaOrganization(taOrganization);
    if (addTaOrganization == 1) {
        return "true";
    }
```

```
        return "新增失败";
}
```

这里转换数组时用到了Arrays.stream方法，将数组转换成数据流，然后转换成map，最后将map中的值以","相隔拼接起来。转换成map时使用了匿名函数->，这是Java8中的Lambda表达式。

Lambda表达式的语法如下：

```
(parameters) -> expression 或 (parameters) ->{ statements; }
```

它有如下四个特性：

- 可选类型声明：不需要声明参数类型，编译器可以统一识别参数值。
- 可选的参数圆括号：一个参数无须定义圆括号，但多个参数需要定义圆括号。
- 可选的大括号：如果主体包含了一个语句，就不需要使用大括号。
- 可选的返回关键字：如果主体只有一个表达式返回值，则编译器会自动返回值，大括号需要指定表达式以返回一个数值。

Lambda表达式只能引用标记了final的外层局部变量，不能在Lambda内部修改定义在域外的局部变量，否则会编译错误。以下为Lambda表达式的简单示例。

```
// 1. 不需要参数,返回值为 2
() -> 2
// 2. 接收一个参数(数字类型)，返回其5倍的值
x -> 5 * x
// 3. 接受2个参数(数字)，并返回它们的差值
(x, y) -> x - y
```

Lambda表达式主要用来定义行内执行的方法类型接口，免去了使用匿名方法的麻烦，并且给予Java简单但又强大的函数化的编程能力。

5.4.6 其他管理

除了以上模块之外，还有角色管理、组织类型管理和按键管理。这三部分管理内容与前面描述的模块基本类似，没有比较特殊的内容，所以这里就不再详述，读者可以自行参考本书配套的源码。

5.5 项目总结

本章实现了一个用户权限管理系统，后端使用的主要技术框架有Spring、Spring MVC、MyBatis和Shrio，前端使用的技术主要有Thymeleaf，数据库使用的是PostgreSQL。本章读者要掌握的重点是Shrio框架和PostgreSQL。知道如何通过Spring Boot整合这些框架并做相应配置，以及如何在实际开发过程中使用这个框架。学习时可参考各个框架官方提供的API手册，同时要回顾前面章节的相关内容，以便理解和记忆。

本章的重点技术：

- Shrio框架：Shrio授权认证的原理和架构，如何使用该框架完成认证和授权。
- PostgreSQL数据库：PostgreSQL数据库的下载与安装，主要的数据特性，客户端的使用方法以及在Spring Boot项目中如何使用PostgreSQL。
- Random类：Random随机数生成类的主要方法。
- Lambda表达式：Java 8及以后版本的新特性Lambda表达式的使用方法。

第 6 章

使用小程序上报用户信息

在第5章中,我们介绍并使用Shrio、PostgreSQL以及Thymeleaf完成用户权限管理系统功能的开发与实现。本章将实现一个使用小程序上报用户信息的项目。整个项目包含客户端和后端两大部分:客户端部分为小程序代码,使用uni-app开发;后端部分为后台服务内容,包括小程序用户登录、接口鉴权、使用接口提交信息、获取信息等。

本章主要涉及的知识点有:

- 如何使用Spring Boot集成JWT框架。
- 如何使用Spring Boot集成MyBatis-Plus框架。
- 如何使用MyBatis-Plus框架实现增、删、改、查。
- 如何使用JWT进行认证。
- 后端如何实现小程序登录。

6.1 项目技术选型

本章涉及的小程序上报信息的后台服务,主要实现与小程序官方接口对接进行小程序登录、信息获取以及信息提交功能。业务代码选用的框架是Spring MVC、MyBatis-Plus,权限认证框架采用JWT结合Spring Security,数据库使用MySQL,后台服务无页面逻辑,小程序客户端使用的JS运行框架为Node.js,开发框架为基于Vue的uni-app框架,工具采用HBuilderX。

6.1.1 MyBatis-Plus框架

1. 概述

MyBatis-Plus简称MP,是MyBatis的增强工具,基于MyBatis只做增强不做改变,所以有了MyBatis基础之后使用MyBatis-Plus只会事半功倍。MyBatis-Plus由一个叫苞米豆的组织开发,整个框架使用Apache 2.0开源协议,源码在Gitee和GitHub上,Gitee上对应的地址为

https://gitee.com/baomidou/mybatis-plus，GitHub对应的地址为https://github.com/baomidou/mybatis-plus。目前MyBatis-Plus已经发展到3.X系列，本章所使用的版本为3.5.1。

MyBatis-Plus在基于MyBatis的基础上增加了诸多特性，官网上给出了它的如下特性：

（1）无侵入

只做增强不做改变，引入它不会对现有工程产生影响。

（2）损耗小

启动即会自动注入基本 CURD，性能基本无损耗，直接面向对象操作。

（3）强大的CRUD操作

内置通用Mapper、通用Service，仅仅通过少量配置即可实现单表大部分CRUD操作，更有强大的条件构造器满足各类使用需求。

（4）支持Lambda形式调用

通过Lambda表达式方便地编写各类查询条件，无须再担心字段写错的问题。

（5）支持主键自动生成

支持多达4种主键策略（内含分布式唯一ID生成器Sequence），可自由配置，完美解决主键问题。

（6）支持ActiveRecord模式

支持ActiveRecord形式的调用，实体类只需继承Model类即可进行强大的CRUD操作。

（7）支持自定义全局通用操作

支持全局通用方法注入。

（8）内置代码生成器

采用代码或者Maven插件可快速生成Mapper、Model、Service、Controller层代码，支持模板引擎，更有超多自定义配置等你来使用。

（9）内置分页插件

基于MyBatis物理分页，开发者无须关心具体操作，配置好插件之后，写分页等同于普通List查询。

（10）分页插件支持多种数据库

支持MySQL、MariaDB、Oracle、DB2、H2、HSQL、SQLite、PostgreSQL、SQL Server等多种数据库。

（11）内置性能分析插件

可输出SQL语句以及其执行时间，建议开发测试时启用该功能，能快速揪出慢查询。

（12）内置全局拦截插件

提供全表delete、update操作智能分析阻断，也可自定义拦截规则，预防误操作。

2. 框架结构

图6.1是MyBatis-Plus官方给出的架构图。这个架构图非常简单，但总体已经描述清楚了MyBatis-Plus的组成内容以及作用场景。如果了解了MyBatis框架，再来看这个图就比较容易理解。关于MyBatis的介绍和使用可以参考本书第2章内容。

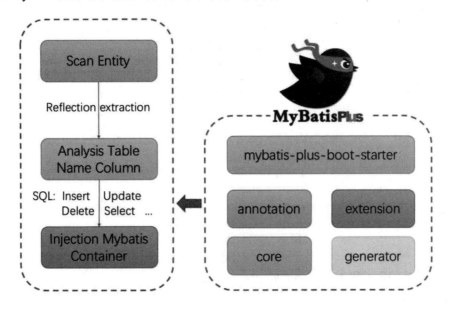

图6.1 MyBatis-Plus架构图

MyBatis-Plus包含启动器、注解、核心库、生成器以及扩展部分。

- 启动器：mybatis-plus-boot-starter，用于Spring Boot，使得Spring可以自动扫描MyBatis-plus的配置内容，初始化MyBatis-Plus的属性。
- 注解：annotation，是MyBatis-Plus支持的注解。通过这些注解，设置实体类与数据库表之间的对应关系，以及简化SQL操作，降低代码量，提高开发效率。
- 扩展：extension，扩展部分包括MyBatis-Plus的增强支持，例如对ActiveRecord的支持、通用枚举、字段类型处理器、SQL注入器、数据安全保护、多数据源等功能。
- 核心部件：core，即实现通用的CRUD功能，包含BaseMapper等，使用时直接自定义Mapper继承该BaseMapper即可获得。
- 生成器：generator，生成实体类、Mapper、XML、Service、Controller等通用代码，支持配置数据源、自定义包名和路径。这部分功能在3.5.1版本之后做了较大改变。

除以上部分外，MyBatis-Plus还有插件部分。例如分页插件、乐观锁插件、多租户插件、防全表更新与删除插件与动态表名插件。

3. 注解

MyBatis-Plus常用的注解如表6.1所示。

表6.1　MyBatis-Plus常用注解

注 解 名	使用位置	描 述
@TableName	实体类	表名注解，标识实体类对应的数据库表名
@TableId	主键字段	主键注解
@TableField	非主键字段	字段注解
@Version	字段	乐观锁注解
@EnumValue	枚举字段	普通枚举类注解
@TableLogic	逻辑删除字段	表字段逻辑处理注解
@KeySequence	Oracle数据库，主键字段	序列主键策略
@OrderBy	排序字段	内置SQL默认指定排序，优先级低于wrapper条件查询

下面来看每个注解的详细使用方法。

（1）@TableName

表名注解，标识实体类对应的表。该注解支持的属性如表6.2所示。

表6.2　@TableName注解属性

属　性	类　型	必须指定	默 认 值	描　述
value	String	否	""	表名
schema	String	否	""	schema
keepGlobalPrefix	boolean	否	false	是否保持使用全局的tablePrefix的值（当全局tablePrefix生效时）
resultMap	String	否	""	XML中resultMap的id（用于满足特定类型的实体类对象绑定）
autoResultMap	boolean	否	false	是否自动构建resultMap并使用（如果设置resultMap则不会进行resultMap的自动构建与注入）
excludeProperty	String[]	否	{}	需要排除的属性名

（2）@TableId

主键注解，使用在实体类主键字段上，用来标注实体类的主键。它支持的属性如表6.3所示。

表6.3　@TableId的属性

属　性	类　型	必须指定	默 认 值	描　述
value	String	否	""	主键字段名
type	Enum	否	IdType.NONE	指定主键类型

IdType枚举类型为主键生成类型。@TableId中的type可选值为枚举类型。表6.4是常用的主键生成类型。

表6.4　IdType可选值

值	描述
AUTO	数据库ID自增
NONE	无状态，该类型为未设置主键类型
INPUT	插入前自行设置主键值
ASSIGN_ID	支持3.3.0及以后版本，分配ID，主键的类型为Number(Long 或Integer)或String，使用接口IdentifierGenerator的方法nextId，默认实现类为DefaultIdentifierGenerator雪花算法，支持自定义
ASSIGN_UUID	支持 3.3.0 及 以 后 版 本，分配 UUID，主 键 的 类 型 为 String，默 认 使 用 接 口 IdentifierGenerator的方法nextUUID，可以自定义

（3）@TableField

该注解用于字段注解，支持的属性如表6.5所示。

表6.5　@TableField注解属性

属性	类型	必须指定	默认值	描述
value	String	否	""	数据库字段名
exist	boolean	否	true	是否为数据库表字段
condition	String	否	""	字段where实体查询比较条件，有值设置则按设置的值为准，没有则为默认全局的%s=#{%s}，参考(opens new window)
update	String	否	""	字段update set部分注入，例如：当在version字段上注解update= "%s+1"表示更新时，set version=version+1（该属性优先级高于el属性）
insertStrategy	Enum	否	FieldStrategy.DEFAULT	举例：NOT_NULL insert into table_a(<if test= "columnProperty != null"> column</if>) values (<if test= "columnProperty != null"> #{columnProperty} </if>)
updateStrategy	Enum	否	FieldStrategy.DEFAULT	举例：IGNORED update table_a set column= #{columnProperty}

(续表)

属性	类型	必须指定	默认值	描述
whereStrategy	Enum	否	FieldStrategy.DEFAULT	举例：NOT_EMPTY where \<if test="columnProperty != null and columnProperty!=""\> column=#{columnProperty}\</if\>
fill	Enum	否	FieldFill.DEFAULT	字段自动填充策略
select	boolean	否	true	是否进行select查询
keepGlobalFormat	boolean	否	False	是否保持使用全局的format进行处理
jdbcType	JdbcType	否	JdbcType.UNDEFINED	JDBC类型（该默认值不代表会按照该值生效）
typeHandler	Class\<? extends TypeHandler\>	否	UnknownTypeHandler.class	类型处理器（该默认值不代表会按照该值生效）
numericScale	String	否	""	指定小数点后保留的位数

以上属性中，关于SQL操作的几个属性，例如insertStrategy、updateStrategy等，使用到了FieldStrategy枚举类型，表示的是执行SQL操作时对该字段采用的策略。支持的策略如表6.6所示。

表6.6　FieldStrategy策略选值

值	描述
IGNORED	忽略判断
NOT_NULL	非NULL判断
NOT_EMPTY	非空判断（只针对字符串类型字段，其他类型字段依然为非NULL判断）
DEFAULT	追随全局配置
NEVER	不加入SQL

fill属性用到了FieldFill枚举类型，表示的是执行SQL操作时该字段是否自动填充。FieldFill的优先级高于FieldStrategy设置，它支持的场景如表6.7所示。

表6.7　FieldFill可选值

值	描述
DEFAULT	默认不处理
INSERT	插入时填充字段
UPDATE	更新时填充字段
INSERT_UPDATE	插入和更新时填充字段

（4）@Version

这是乐观锁注解，标记在版本字段上，版本字段的类型可以是Long、Integer、Date、TimeStamp或LocalDateTime。

（5）@EnumValue

注解在枚举字段上的普通枚举注解。

（6）@TableLogic

标记在实体类的逻辑删除字段上，如果全局数据库表的逻辑删除字段均相同，也可以在全局中配置。

（7）@KeySequence

序列主键策略，在Oracle中使用。

（8）@OrderBy

排序注解，优先级低于Wrapper条件查询，在执行MybatisPlus的自带方法时自动拼接上OrderBy语句，自带方法包含selectList()、Page()等。

4. 核心功能

MyBatis-Plus核心功能包含通用的CRUD操作、条件构造器、主键策略、自定义ID生成器等。

（1）CRUD接口

MyBatis-Plus中的CRUD接口包含Mapper和Service两种。Mapper CRUD接口封装在BaseMapper中，源码如下：

```java
public interface BaseMapper<T> extends Mapper<T> {
    // 插入一条记录
    int insert(T entity);
    // 根据 ID 删除
    int deleteById(Serializable id);
    //根据 columnMap 条件删除记录
    int deleteByMap(@Param(Constants.COLUMN_MAP) Map<String, Object> columnMap);
    // 根据 entity 条件删除记录
    int delete(@Param(Constants.WRAPPER) Wrapper<T> wrapper);
    //删除（根据ID 批量删除）
    int deleteBatchIds(@Param(Constants.COLLECTION) Collection<? extends Serializable> idList);
    //根据 ID 修改
    int updateById(@Param(Constants.ENTITY) T entity);
    // 根据 whereEntity 条件更新记录
    int update(@Param(Constants.ENTITY) T entity, @Param(Constants.WRAPPER) Wrapper<T> updateWrapper);
    // 根据 ID 查询
    T selectById(Serializable id);
    //查询（根据ID 批量查询）
    List<T> selectBatchIds(@Param(Constants.COLLECTION) Collection<? extends Serializable> idList);
    // 查询（根据 columnMap 条件）
    List<T> selectByMap(@Param(Constants.COLUMN_MAP) Map<String, Object> columnMap);
    // 根据 entity 条件查询一条记录
```

```
    T selectOne(@Param(Constants.WRAPPER) Wrapper<T> queryWrapper);
    //根据 Wrapper 条件查询总记录数
    Integer selectCount(@Param(Constants.WRAPPER) Wrapper<T> queryWrapper);
    // 根据 entity 条件查询全部记录
    List<T> selectList(@Param(Constants.WRAPPER) Wrapper<T> queryWrapper);
    // 根据 Wrapper 条件查询全部记录
    List<Map<String, Object>> selectMaps(@Param(Constants.WRAPPER) Wrapper<T> queryWrapper);
    //根据 Wrapper 条件查询全部记录
    List<Object> selectObjs(@Param(Constants.WRAPPER) Wrapper<T> queryWrapper);
    //根据 entity 条件查询全部记录(并翻页)
    <E extends IPage<T>> E selectPage(E page, @Param(Constants.WRAPPER) Wrapper<T> queryWrapper);
    //根据 Wrapper 条件查询全部记录(并翻页)
    <E extends IPage<Map<String, Object>>> E selectMapsPage(E page, @Param(Constants.WRAPPER) Wrapper<T> queryWrapper);
}
```

Mapper层的CRUD涵盖了Insert、Delete、Update和Select操作。其中Select还支持分页查询。泛型T为任意实体对象,参数Serializable 为任意类型的主键。对象Wrapper为条件构造器。

除了Mapper CRUD接口外,MyBatis-Plus还提供了Service CRUD接口,使用时自定义Service接口继承IService接口,并使用ServiceImpl实现类继承ServiceImpl即可。为了与Mapper的CRUD作区分,Service CRUD支持的接口命名方式为Save、Update、Remove、Get、List和Page等。常用的部分方法源码如下:

```
// 插入一条记录(选择字段,策略插入)
boolean save(T entity);
// 插入(批量)
boolean saveBatch(Collection<T> entityList);
// 插入(批量)
boolean saveBatch(Collection<T> entityList, int batchSize);
// TableId 注解存在更新记录,则插入一条记录
boolean saveOrUpdate(T entity);
// 根据updateWrapper尝试更新,否继续执行saveOrUpdate(T)方法
boolean saveOrUpdate(T entity, Wrapper<T> updateWrapper);
// 批量修改插入
boolean saveOrUpdateBatch(Collection<T> entityList);
// 批量修改插入
boolean saveOrUpdateBatch(Collection<T> entityList, int batchSize);
// 根据 entity 条件删除记录
boolean remove(Wrapper<T> queryWrapper);
// 根据 ID 删除
boolean removeById(Serializable id);
// 根据 columnMap 条件删除记录
boolean removeByMap(Map<String, Object> columnMap);
// 删除(根据ID 批量删除)
boolean removeByIds(Collection<? extends Serializable> idList);
// 根据 UpdateWrapper 条件更新记录,需要设置sqlset
boolean update(Wrapper<T> updateWrapper);
// 根据 whereWrapper 条件更新记录
```

```java
    boolean update(T updateEntity, Wrapper<T> whereWrapper);
    // 根据 ID 选择修改
    boolean updateById(T entity);
    // 根据ID 批量更新
    boolean updateBatchById(Collection<T> entityList);
    // 根据ID 批量更新
    boolean updateBatchById(Collection<T> entityList, int batchSize);
    // 根据 ID 查询
    T getById(Serializable id);
    // 根据Wrapper查询一条记录。如果有多个结果集，会抛出异常，随机取一条加上限制条件 wrapper.last("LIMIT 1")
    T getOne(Wrapper<T> queryWrapper);
    // 根据 Wrapper查询一条记录
    T getOne(Wrapper<T> queryWrapper, boolean throwEx);
    // 根据 Wrapper查询一条记录
    Map<String, Object> getMap(Wrapper<T> queryWrapper);
    // 根据 Wrapper查询一条记录
    <V> V getObj(Wrapper<T> queryWrapper, Function<? super Object, V> mapper);
    // 查询所有
    List<T> list();
    // 查询列表
    List<T> list(Wrapper<T> queryWrapper);
    // 查询（根据ID 批量查询)
    Collection<T> listByIds(Collection<? extends Serializable> idList);
    // 查询（根据 columnMap 条件）
    Collection<T> listByMap(Map<String, Object> columnMap);
    // 查询所有列表
    List<Map<String, Object>> listMaps();
    // 查询列表
    List<Map<String, Object>> listMaps(Wrapper<T> queryWrapper);
    // 查询全部记录
    List<Object> listObjs();
    // 查询全部记录
    <V> List<V> listObjs(Function<? super Object, V> mapper);
    // 根据 Wrapper 条件查询全部记录
    List<Object> listObjs(Wrapper<T> queryWrapper);
    // 根据 Wrapper 条件查询全部记录
    <V> List<V> listObjs(Wrapper<T> queryWrapper, Function<? super Object, V> mapper);
    // 无条件分页查询
    IPage<T> page(IPage<T> page);
    // 条件分页查询
    IPage<T> page(IPage<T> page, Wrapper<T> queryWrapper);
    // 无条件分页查询
    IPage<Map<String, Object>> pageMaps(IPage<T> page);
    // 条件分页查询
    IPage<Map<String, Object>> pageMaps(IPage<T> page, Wrapper<T> queryWrapper);
    // 查询总记录数
    int count();
    // 根据 Wrapper 条件查询总记录数
    int count(Wrapper<T> queryWrapper);
```

除了以上通用的CRUD操作之外，MyBatis-Plus还扩展支持了ActiveRecord模式。实体类只需继承 Model 类即可进行强大的 CRUD 操作。

（2）条件构造器

MyBatis-Plus提供的条件构造器如图6.2所示。

图6.2　条件构造器

主要通过QueryWrapper和UpdateWrapper进行条件构造，这两个和LambdaQueryWrapper、LambdaUpdateWrapper 差不多是等价的，只不过后者采用了JDK8提供的Lambda语法，使用起来更便捷。

AbstractWrapper构造器以及它的子类和实现类支持的方法如表6.8所示。

表6.8　构造器支持的方法

方法名	说明	用法实例	等价SQL
allEq(Map<R, V> params)	全部等于	allEq({id:1,name: "老王",age:null})	id = 1 and name = '老王' and age is null
eq(R column, Object val)	等于（=）	eq("name", "老王")	name = '老王'
ne(R column, Object val)	不等于（<>）	ne("name", "老王")	name <> '老王'
gt(R column, Object val)	大于（>）	gt("age", 18)	age > 18
ge(R column, Object val)	大于等于（>=）	ge("age","18")	age >= 18
lt(R column, Object val)	小于（<）	lt("age","18")	age < 18
le(R column, Object val)	小于等于（<=）	le("age","18")	age <= 18
between(R column, Object val1, Object val2)	BETWEEN值1 AND值2	between("age","18","25")	age BETWEEN 18 AND 25
notBetween(R column, Object val1, Object val2)	NOT BETWEEN值1 AND值2	notBetween("age","18","25")	age NOT BETWEEN 18 AND 25
like(R column, Object val)	LIKE '%值%'	like("name","王")	like '%王%'
notLike(R column, Object val)	NOT LIKE '%值%'	notLike("name","王")	not like '%王%'
likeLeft(R column, Object val)	LIKE '%值'	likeLeft("name","王")	like '%王'
likeRight(R column, Object val)	LIKE '值%'	likeRight("name","王")	like '王%'

（续表）

方法名	说明	用法实例	等价SQL
isNull(R column)	字段IS NULL	isNull("name")	name IS NULL
isNotNull(R column)	字段IS NOT NULL	isNotNull("name")	name IS NOT NULL
in(R column, Collection<?> value)	字段IN (value.get(0), value.get(1), …)	in("age",{1,2,3})	age IN (1,2,3)
notIn(R column, Collection<?> value)	字段NOT IN (value.get(0), value.get(1), …)	notIn("age",{1,2,3})	age NOT IN (1,2,3)
inSql(R column, String inValue)	字段IN（SQL语句）	inSql("id","select id from user")	id IN (select id from user)
notInSql(R column, String inValue)	字段NOT IN（SQL语句）	notInSql("id", "select id from user where id > 2")	id NOT IN (select id from user where id > 2
groupBy(R… columns)	分组：GROUP BY 字段, …	groupBy("id", "name")	GROUP BY id,name
orderByAsc(R… columns)	排序：ORDER BY 字段, … ASC	orderByAsc("id", "name")	ORDER BY id ASC,name ASC
orderByDesc(R… columns)	排序：ORDER BY 字段, … DESC	orderByDesc("id", "name")	ORDER BY id DESC,name DESC
orderBy(boolean condition, boolean isAsc, R… columns)	ORDER BY 字段, …	orderBy(true,true,"id", "name")	ORDER BY id ASC,name ASC
having(String sqlHaving, Object… params)	HAVING（SQL语句）	having("sum(age)>{0}", "25")	HAVING sum(age)>25
or()	拼接OR	eq("id",1).or().eq("age",25)	id = 1 OR age = 25
and(Consumer consumer)	AND嵌套	and(i->i.eq("id",1).ne("age",18))	id = 1 AND age <> 25
nested(Consumer consumer)	正常嵌套 不带 AND 或者 OR	nested(i->i.eq("id",1).ne("age", 18))	id = 1 AND age <> 25
apply(String applySql, Object… params)	拼接SQL（不会有SQL注入风险）	apply("age>{0}","25 or 1=1")	age >'25 or 1=1'
last(String lastSql)	拼接到SQL的最后，多次调用以最后一次为准（有SQL注入的风险）	last("limit 1")	limit 1
exists(String existsSql)	拼接EXISTS（SQL语句）	exists("select id from user where age = 1")	EXISTS (select id from user where age = 1)
notExists(String notExistsSql)	拼接 NOT EXISTS（SQL语句）	notExists("select id from user where age = 1")	NOT EXISTS (select id from user where age = 1)

QueryWrapper和LambdaQueryWrapper支持select方法实现查询字段过滤。

```
select(String... sqlSelect)
select(Predicate<TableFieldInfo> predicate)
select(Class<T> entityClass, Predicate<TableFieldInfo> predicate)
```

例如：

```
select("id", "name", "age")
select(i -> i.getProperty().startsWith("test"))
```

UpdateWrapper和LambdaUpdateWrapper支持set方法实现SQL字段设置。

```
set(String column, Object val)
set(boolean condition, String column, Object val)
setSql(String sql)
```

例如：

```
set("name", "老李头")
setSql("name = '老李头'")
```

除了以上构造器提供的方法外，MyBatis-Plus 3.0.7及以后版本也支持使用Wrapper自定义SQL。自定义SQL时，可以使用注解形式。代码如下：

```
@Select("select * from mysql_data ${ew.customSqlSegment}")
List<MysqlData> getAll(@Param(Constants.WRAPPER) Wrapper wrapper);
```

也可以使用XML形式，参见第2章MyBatis部分内容。

```
List<MysqlData> getAll(Wrapper ew);
//以下为XML文件内容
<select id="getAll" resultType="MysqlData">
    SELECT * FROM mysql_data ${ew.customSqlSegment}
</select>
```

5. 扩展内容

扩展内容中包含ActiveRecord、通用枚举、字段类型处理器、SQL注入器、数据安全保护、多数据源等功能。

ActiveRecord模式使用的前提条件有两个：

（1）实体类需要继承Model类。

（2）对应实体的BaseMapper已注入。

继承Model类时，传入对应的实体泛型。

```
class User extends Model<User>{
    // fields...
}
```

然后就可以直接使用CRUD方法调用API完成CRUD操作。

```
User user = new User();
user.insert();
user.selectAll();
user.updateById();
user.deleteById();
// ...
```

关于其他扩展功能，读者可以自行去MyBatis-Plus官网查阅相关文档。

6. 插件

目前MyBatis-Plus已有的插件如下：

- 自动分页：PaginationInnerInterceptor。
- 多租户：TenantLineInnerInterceptor。
- 动态表名：DynamicTableNameInnerInterceptor。
- 乐观锁：OptimisticLockerInnerInterceptor。
- SQL性能规范：IllegalSQLInnerInterceptor。
- 防止全表更新与删除：BlockAttackInnerInterceptor。

使用多个功能时需要注意顺序关系，建议使用如下顺序：多租户→动态表名→分页→乐观锁→SQL性能规范→防止全表更新与删除。

在Spring Boot中使用插件时，需要注入对应的bean。代码如下：

```java
@Configuration
@MapperScan("scan.your.mapper.package")
public class MybatisPlusConfig {
    /**
     * 新的分页插件，一缓和二缓遵循MyBatis的规则，需要设置 MybatisConfiguration
#useDeprecatedExecutor = false 避免缓存出现问题(该属性会在旧插件移除后一同移除)
     */
    @Bean
    public MybatisPlusInterceptor mybatisPlusInterceptor() {
        MybatisPlusInterceptor interceptor = new MybatisPlusInterceptor();
        interceptor.addInnerInterceptor(new PaginationInnerInterceptor(DbType.H2));
        return interceptor;
    }

    @Bean
    public ConfigurationCustomizer configurationCustomizer() {
        return configuration -> configuration.setUseDeprecatedExecutor(false);
    }
}
```

其他插件参考此方式即可。

6.1.2 JWT

JWT（JSON Web Token）是一个开放标准（RFC 7519），它定义了一种紧凑的、自包含的方式，用作JSON对象在各方之间安全地传输信息。由于信息已被数字认证，所以是正确且可信赖的。签发JWT时，可以使用HMAC算法加密，也可以使用RSA或ECDSA公私钥加密方式。换句话说，JWT提供了一种认证机制，使得后台服务接口认可请求是来自于可信赖的客户端的。

JWT的常见使用场景如下：

- 认证：这是JWT最常见的使用场景。一旦用户登录系统后，后续的每一个请求都需要携带JWT才能访问路由、服务以及资源。目前，由于JWT的轻巧简便以及它对跨域的支持，因此多用于单点登录。

- 信息交换：在两个主体之间加密传递信息时，JWT是一个非常棒的方式。由于JWT能够使用公私钥对进行签发，所以能够确定信息发送者的身份。另外，由于签名是通过头部和有效荷载信息来计算的，所以可以验证发送的内容是否被篡改。

JWT的结构非常紧凑，它由三部分组成，每个部分之间以"."隔开，这三部分分别是：

- 头部。
- 有效荷载。
- 签名。

所以一个典型的JWT如下所示。

```
xxxxx.yyyyy.zzzzz
```

1. 头部

头部一般由两部分组成：token类型和所使用的算法。代码如下：

```
{
  "alg": "HS256",
  "typ": "JWT"
}
```

发送请求时，将上述信息进行Base64Url编码作为JWT的第一部分。

2. 有效荷载

JWT的第二部分是有效荷载，包含的内容是断言，即关于主体（一般是用户）以及附属数据的声明。断言分为三种：

注册断言（Register claims）：这是预定义的断言集合，虽然不是必须的，但推荐使用。它提供一系列有用的、互相操作的断言集合。例如iss（发行者）、exp（过期时间）、sub（主体）、aud（受众）等。在JWT中注册断言的长度是固定的3个字符长度。

公共断言（Public claims）：公共断言根据JWT使用者的意愿来定义。但是为了避免冲突，所有的断言必须在JWT注册表中定义，或者在URI中定义，这个URI包含一个避免冲突的命名空间。

私有断言（Private claims）：私有断言是客户端断言，用来在不同参与者之间分享信息。

所以，一个有效荷载的例子如下：

```
{
  "sub": "1234567890",
  "name": "John Doe",
  "admin": true
}
```

其中sub是注册断言，name是公共断言，admin是私有断言。以上内容经过Base64Url编码后作为JWT的第二部分。

3. 签名

生成签名前，需要准备编码后的头部信息和有效荷载信息、一个密钥以及头部标明的算法。该密钥仅保存在服务器中，保证不能让其他用户知道。例如，使用HMAC SHA256算法生成签名：

```
HMACSHA256(
  base64UrlEncode(header) + "." +
  base64UrlEncode(payload),
  secret)
```

签名能够用来验证信息是否被修改。另外，如果签名是以私钥签发的，通过签名也能验证该JWT的发送者。

综上所述，头部、有效荷载、签名三者结合在一起即可组成一个有效的JWT，如图6.3所示。

```
eyJhbGciOiJIUzI1NiIsInR5cCI6IkpXVCJ9.
eyJzdWIiOiIxMjM0NTY3ODkwIiwibmFtZSI6IkpvaG4
gRG9lIiwiaXNTb2NpYWwiOnRydWV9.
4pcPyMD09olPSyXnrXCjTwXyr4BsezdI1AVTmud2fU4
```

图6.3 JWT示例

了解了JWT的使用场景和组成结构之后，我们来看一下JWT是如何实现认证的。一个简单的JWT认证流程如图6.4所示。

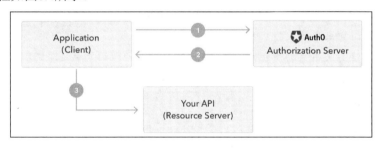

图6.4 JWT认证流程

其中，Authorization Server是认证服务端，Application是客户端，Your API是资源服务端。客户端先向认证服务端发送认证请求，一般是先登录，登录通过后签发一个JWT返回到客户端，客户端收到JWT后将它存储到客户端，然后带着这个JWT向其他资源服务端发送请求或者请求API接口。服务端接收到JWT后根据前两部分和所存储的密钥计算出一个签名值，与传递过来的JWT签名部分作比较，如果两者匹配，代表认证通过。

需要注意的是，默认情况下JWT是未加密的，任何人都可以解读其内容，因此一些敏感信息不要存放在此，以防信息泄露。

JWT有如下优点：

- 具有JSON格式的通用性，所以JWT可以跨语言支持，比如Java、JavaScript、PHP、Node等。
- 可以利用Payload存储一些非敏感的信息。
- JWT结构简单，字节占用小，便于传输。
- 不需要在服务端保存会话信息，易于应用的扩展。

6.1.3 HbuilderX简介

HBuilderX（简称HX）是DCloud（数字天堂）推出的一款支持HTML5的Web开发IDE。"快"

是HBuilderX的最大优势，通过完整的语法提示和代码输入法、代码块等大幅提升HTML、JS、CSS的开发效率。是目前前端开发工具的主流选择。

它的官网地址为https://www.dcloud.io/hbuilderx.html，在这可以进行下载安装。它的产品文档网址为https://hx.dcloud.net.cn/，在这可以找到对应的安装手册和使用教程。

官网上介绍的HBuilderX的优点如下：

- 轻巧：仅10余兆字节的绿色发行包。
- 极速：不管是启动速度、大文档打开速度还是编码提示，都极速响应。C++的架构性能远超Java或Electron架构。
- vue开发强化：HX对vue做了大量优化投入，开发体验远超其他开发工具。国外开发工具没有对中国的小程序开发优化，HX可新建uni-app小程序等项目，为国人提供更高效工具。
- markdown利器：HX是唯一一个新建文件默认类型是markdown的编辑器，也是对md支持最强的编辑器。HX为md强化了众多功能。
- 清爽护眼：HX的页面比其他工具更清爽简洁，绿柔主题经过科学的脑疲劳测试，是最适合人眼长期观看的主题页面。
- 强大的语法提示：HX是中国唯一一家拥有自主IDE语法分析引擎的公司，对前端语言提供准确的代码提示和转到定义。
- 高效极客工具：更强大的多光标、智能双击等，让字处理的效率大幅提升。
- 更强的JSON支持：现代JS开发中含有大量JSON结构的写法，HX提供了比其他工具更高效的操作。

另外，HBuilderX还支持插件扩展，它支持Java插件、Node.js插件，并兼容了很多VSCode的插件及代码块。HBuilderX插件市场拥有丰富的插件，对于提升工作效率有极大帮助。插件市场的网址为https://ext.dcloud.net.cn/。

HBuilderX支持绿色免安装，支持多种操作系统。这里我们以Windows10为例，演示它的下载安装过程。

步骤 01 首先到官网下载对应的安装包，目前的版本为V3.4.15，如图6.5所示。

图6.5　HBuilderX下载页面

步骤02　下载完成后直接解压安装包，如图6.6所示。

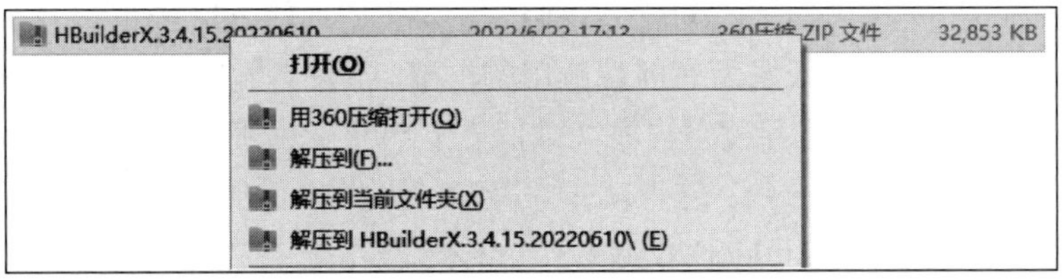

图6.6　解压安装包

解压完成后，目录如图6.7所示。

图6.7　HBuilderX解压目录

步骤03　双击HBuilderX即可运行，为了方便操作，可以将此应用程序发送到桌面。

6.1.4　小程序客户端项目搭建

在进行小程序的项目搭建前我们需要作如下准备：

- 微信平台账号：申请一个微信官方的小程序平台账号，地址为https://mp.weixin.qq.com/，申请完成后，获取小程序的AppId。若仅仅为了学习，这步可以省略，可以使用微信官方提供的测试id，但要注意，该AppId需要与HBuilderX发行时所填的AppId相同。
- 微信开发者工具：微信官方提供的微信开发工具，详情参考6.1.5节。

安装完HBuilderX之后，就可以使用它来创建项目。HBuilderX对前端项目非常友好。我们可以用它创建各种类型的前端项目。对于本项目来说，我们使用uni-app框架开发小程序应用。

uni-app是一个使用Vue.js开发所有前端应用的框架。开发者编写一套代码，可以发布到iOS、Android、Web（响应式）以及各种小程序（微信、支付宝、百度、头条、飞书、QQ、快手、钉钉、淘宝）、快应用等多个平台。这也就大大减少了开发人员的工作量，并实现了多平台项目功能同步。下面我们使用HBuilderX创建一个uni-app应用。

首先在HBuilderX的菜单栏中单击"文件"→"新建"→"项目"命令，如图6.8所示。

图6.8　新建项目

弹出的项目类型选择页面如图6.9所示。

图6.9　选择项目类型

在这个页面中，可以看到HBuilderX支持的各种项目类型，同时也能看出它强大的插件扩展功能。

这里我们选择uni-app，创建完成后，项目会自动生成一些基础的插件、组件和页面，如图6.10所示。我们可以根据自己的需要对项目中的内容进行删减。

小程序客户端开发完成后，可以使用发行功能将小程序发布到小程序开发工具中，这样就可以完成小程序代码的上传，如图6.11所示。

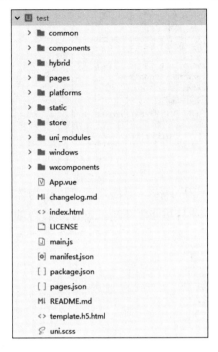

图6.10　新建项目目录

图6.11　发行小程序

发行时，需要填入前面我们申请的小程序官方AppId，如图6.12所示。

图6.12　输入小程序AppId

6.1.5　微信开发者工具

为了帮助开发者简单且高效地开发和调试微信小程序，微信官方提供了微信开发者工具，集成了公众号网页调试和小程序调试两种开发模式。当我们使用HBuilderX开发完uni-app项目后，需要发行到该工具上才能完成小程序代码的编写。

微信开发者工具的下载地址为https://developers.weixin.qq.com/miniprogram/dev/devtools/stable.html。它的简介及使用教程地址为https://developers.weixin.qq.com/miniprogram/dev/devtools/devtools.html。由于篇幅关系，不再演示它的下载与安装过程。微信官方提供的文档讲解得非常详细，读者可以前往官网查看。

小程序代码发行之后，就会自动打开微信开发者工具，如图6.13所示。

```
17:53:29.714
[广告] 17:53:29.739 DCloud 2022新春招聘开启，欢迎前端、Android、C++/QT
[HBuilder] 17:53:29.742  项目 'dmd-miniapp' 开始发布微信小程序...
[HBuilder] 17:53:29.771  项目 'dmd-miniapp' 开始编译...
[HBuilder] 17:53:30.986  小程序各家浏览器内核及自定义组件实现机制存在差异
[HBuilder] 17:53:30.997  正在编译中...
[HBuilder] 17:53:33.353  Browserslist: caniuse-lite is outdated. Ple
[HBuilder] 17:53:33.357  npx browserslist@latest --update-db
[HBuilder] 17:53:47.763  DONE  Build complete.
[HBuilder] 17:53:47.764  项目 'dmd-miniapp' 编译成功。
[HBuilder] 17:53:48.507  项目 'dmd-miniapp'导出微信小程序成功，路径为：
[HBuilder] 17:53:48.517  正在启动微信开发者工具...
[HBuilder] 17:53:49.238  [微信小程序开发者工具] - initialize
```

图6.13　发行日志

打开成功后，我们可以进行小程序代码调试以及代码上传。上传之后可以作为体验版进行体验，正式发布之前需要申请HTTPS证书。由于本章项目重点在于Spring Boot后台项目与小程序的数据交互，所以此处不再详述。

6.1.6　后台服务框架搭建

1. 使用Spring Boot集成MyBatis-Plus

（1）引入依赖

首先引入Maven依赖：

```xml
<dependency>
    <groupId>com.baomidou</groupId>
    <artifactId>mybatis-plus-boot-starter</artifactId>
    <version>${mybatis-plus.version}</version>
</dependency>
```

这里的${mybatis-plus.version}是使用的引用版本号，事先在properties中声明。

```xml
<properties>
    <mybatis-plus.version>3.5.1</mybatis-plus.version>
</properties>
```

然后在Web项目的依赖中引入如下依赖：

```xml
<dependency>
    <groupId>org.mybatis.spring.boot</groupId>
    <artifactId>mybatis-spring-boot-starter</artifactId>
    <version>2.2.2</version>
</dependency>
```

（2）修改Spring Boot配置

接下来在Spring Boot的配置文件中进行配置。本章选择yaml形式。配置的结构如下：

```yaml
mybatis-plus:
  ...
  configuration:
    ...
```

```
global-config:
    ...
    db-config:
        ...
```

- Configuration：大都为MyBatis原生支持的配置，这意味着可以通过MyBatis XML配置文件的形式进行配置。
- global-config：全局配置，其中db-config是全局数据库配置。这些设置也可以通过在实体类上加注解来实现，但为了方便以及提高开发效率，我们通常会在配置文件中做些全局配置。MyBatis-Plus中支持的常用配置如表6.9所示。

表6.9 MyBatis-Plus配置项

配置名称	作用	
configLocation	MyBatis配置文件位置，如果有单独的MyBatis配置，请将其路径配置到configLocation中	
mapperLocations	MyBatis Mapper所对应的XML文件位置，如果在Mapper中有自定义方法（XML中有自定义实现），需要进行该配置，告诉Mapper所对应的XML文件位置	
typeAliasesPackage	MyBaits别名包扫描路径，通过该属性可以给包中的类注册别名，注册后可以在Mapper对应的XML文件中直接使用类名，而不用使用全限定的类名（即在XML中调用的时候不用包含包名）	
typeAliasesSuperType	该配置和typeAliasesPackage一起使用，如果配置了该属性，则仅会扫描路径下以该类作为父类的域对象	
typeHandlersPackage	TypeHandler扫描路径，如果配置了该属性，SqlSessionFactoryBean会把该包下面的类注册为对应的TypeHandler	
checkConfigLocation	仅支持Spring Boot，启动时是否检查MyBatis XML文件的存在，默认不检查	
executorType	仅支持Spring Boot，通过该属性可指定MyBatis的执行器：simple、reuse、batch	
configurationProperties	指定外部化MyBatis Properties配置，通过该配置可以抽离配置，实现不同环境的配置部署	
Configuration	mapUnderscoreToCamelCase	是否开启自动驼峰命名规则（camel case）映射，即从经典数据库列名A_COLUMN（下划线命名）到经典Java属性名aColumn（驼峰命名）的类似映射
	defaultEnumTypeHandler	默认枚举处理类，如果配置了该属性，枚举将统一使用指定处理器进行处理
	aggressiveLazyLoading	当设置为true的时候，懒加载的对象可能被任何懒属性全部加载，否则每个属性都按需加载。需要和lazyLoadingEnabled一起使用
	autoMappingBehavior	MyBatis自动映射策略，通过该配置可指定MyBatis是否并且如何自动映射数据表字段与对象的属性，总共有3种可选值：NONE、PARTIAL、FULL

（续表）

配置名称	作用		
Configuration	autoMappingUnknownColumnBehavior	MyBatis自动映射时未知列或未知属性处理策略，通过该配置可指定MyBatis在自动映射过程中遇到未知列或者未知属性时如何处理，总共有3种可选值：NONE、WARNING、FAILING	
	localCacheScope	MyBatis一级缓存，默认为SESSION	
	cacheEnabled	开启MyBatis二级缓存，默认为true	
	callSettersOnNulls	指定当结果集中值为null的时候是否调用映射对象的Setter（Map对象时为put）方法，通常运用于有Map.keySet()依赖或null值初始化的情况	
	configurationFactory	从3.2.3版本开始，用来指定一个提供Configuration实例的工厂类。该工厂生产的实例将用来加载已经被反序列化对象的懒加载属性值，且必须包含一个签名方法static Configuration getConfiguration()	
	logImpl	打印日志	
GlobalConfig	banner	是否在控制台打印Mybatis-Plus的LOGO	
	enableSqlRunner	是否初始化SqlRunner	
	sqlInjector	SQL注入器	
	superMapperClass	通用Mapper父类	
	metaObjectHandler	元对象字段填充控制器	
	identifierGenerator	从3.3.0版本开始为Id生成器	
	dbConfig	idType	全局默认主键类型
		tablePrefix	表名前缀
		schema	即数据库中的schema
		columnFormat	从3.1.1版本开始为字段format，例：%s（对主键无效）
		propertyFormat	从3.3.0版本开始为entity的字段(property)的format，只有在column as property这种情况下生效例：%s（对主键无效）
		tableUnderline	表名是否使用驼峰转下划线命名，只对表名生效
		capitalMode	大写命名，对表名和字段名均生效
		keyGenerator	表主键生成器
		logicDeleteField	全局的entity的逻辑删除字段属性名
		logicDeleteValue	逻辑已删除值
		logicNotDeleteValue	逻辑未删除值
		insertStrategy	从3.1.2版本开始，字段验证策略之insert，即insert时的字段验证策略

(续表)

配置名称	作用		
GlobalConfig	dbConfig	updateStrategy	从3.1.2版本开始，字段验证策略之update，即update时的字段验证策略
		selectStrategy	从3.1.2版本开始，字段验证策略之select，即select时的字段验证策略，wrapper根据内部entity生成的where条件

本章的MyBatis-Plus配置如下：

```yaml
##MyBatis配置
mybatis-plus:
  # XML扫描，多个目录用逗号或者分号分隔（告诉Mapper所对应的XML文件位置）
  mapper-locations: classpath*:mappings/*.xml
  # 以下配置均有默认值，可以不设置
  global-config:
    db-config:
      #主键类型 AUTO:"数据库ID自增" INPUT:"用户输入ID",ID_WORKER:"全局唯一ID（数字类型唯一ID)", UUID:"全局唯一ID UUID";
      id-type: assign_id
      #字段策略 IGNORED:"忽略判断" NOT_NULL:"非NULL判断" NOT_EMPTY:"非空判断"
        #字段策略 IGNORED:"忽略判断" NOT_NULL:"非NULL判断" NOT_EMPTY:"非空判断"
      update-strategy: IGNORED
      #数据库类型
      db-type: MYSQL
      logic-delete-field: isDeleted  # 全局逻辑删除的实体字段名(since 3.3.0)
      logic-delete-value: "1" # 逻辑已删除值(默认为 1)
      logic-not-delete-value: "0" # 逻辑未删除值(默认为 0)
  configuration:
    # 是否开启自动驼峰命名规则映射:从数据库列名到Java属性驼峰命名的类似映射
    map-underscore-to-camel-case: true
    # 返回map时，true:当查询数据为空时字段返回为null, false:当查询数据为空时，字段将被隐藏
    call-setters-on-nulls: true
    # 这个配置会将执行的SQL打印出来，在开发或测试的时候可以用
    log-impl: org.apache.ibatis.logging.stdout.StdOutImpl
```

参数说明：

- mapper-locations：设置XML文件所在位置。一般情况下，使用MyBatis-Plus无须再自定义XML文件，但某些业务复杂的场景仍需要自定义SQL来实现，可能会用到XML。
- id-type：这里采用的是assign_id。
- update-strategy：这里采用的是IGNORED，即忽略判断正常更新。注意这里和字段上的注解取值有区分，因为字段上可以选择追随全局配置以及忽略该字段。
- db-type：这里使用的是MYSQL。
- log-impl：用于开发或者测试，打印SQL执行的日志。

（3）自定义配置类

完成以上配置之后，在项目config目录下创建自定义的MyBatis-Plus配置类，配置分页插件，代码如下：

```
@Configuration
public class MyBatisPlusConfig {
    //分页用
    @Bean
    public PaginationInterceptor paginationInterceptor(){
        PaginationInterceptor paginationInterceptor = new PaginationInterceptor();
        return paginationInterceptor;
    }
}
```

（4）自定义生成器

由于本次项目采用的是自定义主键生成机制，所以需要编写一个Id生成器，重写主键生成方法。

```
@Component
public class CustomerIdGenerator implements IdentifierGenerator {
    @Value("${customer.work}")
    private Long workId;//yml中配置
    @Value("${customer.data}")
    private Long dataCenterId;//yml中配置
    @Override
    public Number nextId(Object entity) {
        // 填充自己的Id生成器，
        Snowflake snowflake = IdUtil.getSnowflake(workId, dataCenterId);//雪花算法
        return snowflake.nextId();
    }
}
```

至此，就完成了Spring Boot中MyBatis-Plus的集成。

2. 使用Spring Boot集成JWT

由于本项目中并不是单独使用JWT，而是结合Spring Security一起使用，所以集成时要结合Spring Security。结合Spring Security时需要注意的是：Spring Security默认的用户密码验证方式无法满足微信小程序这种无密码的认证，所以需要重写一个过滤器来替换UsernamePasswordAuthenticationFilter。当请求发送过来后，首先验证该请求是否为已登录的用户，如果不是，会继续向下进入到Spring Security的工作流程，所以要在用户密码认证过滤器之前加入JWT相关的过滤器。这里可以结合第4章关于Spring Security的介绍。

（1）引入依赖

使用Spring Boot集成JWT，同样地首先需要引入依赖。

```
<dependency>
    <groupId>io.jsonwebtoken</groupId>
```

```
        <artifactId>jjwt</artifactId>
</dependency>
```

（2）修改Spring Boot配置

在配置文件中设置JWT过期时间。

```
jwt:
  expiration: 7200 # 单位s(秒)
```

（3）创建工具类

创建自定义工具类，在这个工具类中定义生成JWT、解析JWT的方法。

```
@Component
public class JwtTokenUtils {
    @Value("${jwt.expiration}")
    private Long expiration;//过期时间
    public SecretKey key;//密钥key
    public JwtTokenUtils() {
        key = Keys.secretKeyFor(SignatureAlgorithm.HS256);
    }
    //获取token中的用户主体信息
    public String getUsernameFromToken(String token) {
        return getClaimFromToken(token, Claims::getSubject);
    }
    //获取JWT token中的断言信息
    public <T> T getClaimFromToken(String token, Function<Claims, T> claimsResolver){
        final Claims claims = getAllClaimsFromToken(token);
        return claimsResolver.apply(claims);
    }
    //获取JWT token中的断言信息
    private Claims getAllClaimsFromToken(String token) {
        return Jwts.parser()
                .setSigningKey(this.key)
                .parseClaimsJws(token)
                .getBody();
    }
    //签发token
    public String generateToken(String openid) {
        Map<String, Object> claims = Maps.newHashMap();
        return doGenerateToken(claims, openid);
    }
    private String doGenerateToken(Map<String, Object> claims, String subject) {
        final Date createdDate = DateTime.now();
        final Date expirationDate = calculateExpirationDate(createdDate);
        return Jwts.builder()
                .setClaims(claims)
                .setSubject(subject)
                .setIssuedAt(createdDate)
                .setExpiration(expirationDate)
                .signWith(this.key)
                .compact();
```

```
    }
    //计算过期时间
    private Date calculateExpirationDate(Date createdDate) {
        return new Date(createdDate.getTime() + expiration * 1000);
    }
}
```

(4)创建Spring Security相关类

由于本章项目涉及小程序的登录,所以这里创建一个用于微信登录的Token类,用于认证。

```
public class WxAppletAuthenticationToken extends AbstractAuthenticationToken {
    private String openid;
    private String sessionKey;
    private String accessToken;
    private String ip;
    private String unionid;
    private String source;//空小程序,app手机客户端
    private String nickname;
    private Integer sex;
    private String headimgurl;
    ...
}
```

然后创建自定义的WxAppletAuthenticationManager,它继承自AuthenticationManager,用于账号登录验证。这里需要重写authenticate()方法。通过该认证方法之后获取账号信息以及权限相关信息,返回上一步自定义的WxAppletAuthenticationToken对象,该对象中包含当前登录用户的信息,然后进入认证结果处理器。

```
...
//获取权限
List<Permission> permissions =
authService.acquirePermission(account.getAccountId());
    List<SimpleGrantedAuthority> authorities = permissions.stream().map(permission ->
new SimpleGrantedAuthority(permission.getPermission())).collect(Collectors.toList());
    return new WxAppletAuthenticationToken(account.getOpenid(),
            account.getSessionKey(), account.getAccessToken(),
            account.getUnionid(), wxAppletAuthenticationToken.getSource(),
            authorities);
```

(5)创建认证结果处理器

认证结果有两种:成功与失败,这两种情况对应的过滤器都要声明。成功过滤器继承自AuthenticationSuccessHandler,重写onAuthenticationSuccess方法,在该方法中签发JWT的token,返回到客户端。失败过滤器继承自AccessDeniedHandler,重写其中的handle方法。

(6)创建过滤器

这里需要定义两个过滤器,一个是用于登录认证的过滤器WxAppletAuthenticationFilter,另一个是用于验证是否登录的过滤器JwtAuthenticationTokenFilter。

请求发送过来后,首先验证是否登录,即通过JwtAuthenticationTokenFilter。该过滤器首先拿到头部的token值,如果token值存在,使用JWT工具类解析出用户唯一识别标识,可以是

账号，也可以是id。这里使用的是unionid，因为笔者使用的小程序绑定了微信开放平台；如果没有绑定的话，可以直接获取小程序用户的openid，同样是唯一识别标识。然后根据唯一识别标识从数据库中获取用户信息，生成之前定义的登录Token对象并放置到Spring容器中，表示该用户已登录。否则，继续向下进入到登录过滤器WxAppletAuthenticationFilter。

```
protected void doFilterInternal(HttpServletRequest request, HttpServletResponse
response, FilterChain filterChain) throws ServletException, IOException {
    log.debug("processing authentication for [{}]", request.getRequestURI());
    //获取token
    String token = request.getHeader(ConstantEnum.AUTHORIZATION.getValue());
    String unionid = null;
    //token存在
    if (token != null) {
        try {
            //得到小程序对应的unionid或openid
            unionid = jwtTokenUtils.getUsernameFromToken(token);
        } catch (IllegalArgumentException e) {
            log.error("an error occurred during getting username from token", e);
            throw new BasicException(ExceptionEnum.JWT_EXCEPTION.customMessage
("an error occurred during getting username from token , token is [%s]", token));
        } catch (ExpiredJwtException e) {
            log.warn("the token is expired and not valid anymore", e);
            throw new BasicException(ExceptionEnum.JWT_EXCEPTION.customMessage
("the token is expired and not valid anymore, token is [%s]", token));
        }catch (SignatureException e) {
            log.warn("JWT signature does not match locally computed signature", e);
            throw new BasicException(ExceptionEnum.JWT_EXCEPTION.customMessage
("JWT signature does not match locally computed signature, token is [%s]", token));
        }
    }else {
        log.warn("couldn't find token string");
    }
    if (unionid != null && SecurityContextHolder.getContext()
.getAuthentication() == null) {
        log.debug("security context was null, so authorizing user");
        Account account = authService.findAccountByUnionid(unionid);
        List<Permission> permissions =
authService.acquirePermission(account.getAccountId());
        List<SimpleGrantedAuthority> authorities = permissions.stream()
.map(permission -> new SimpleGrantedAuthority(permission.getPermission()))
.collect(Collectors.toList());
        log.info("authorized user [{}], setting security context", unionid);
        SecurityContextHolder.getContext().setAuthentication(
            new WxAppletAuthenticationToken(unionid,authorities));
    }
    filterChain.doFilter(request, response);
}
```

登录过滤器中最主要的是attemptAuthentication方法，在该方法中执行登录操作，登录成功后返回自定义的登录对象，然后进入认证管理器。

```java
    //发送请求到微信平台接口,获取登录结果
    WxLoginResultDTO wxLoginResult = restTemplate.getForObject(url,
WxLoginResultDTO.class);
    wxLoginResult.setSource(source);
    if(wxLoginResult.getErrcode() != null
&& !wxLoginResult.getErrcode().equals(0)){//登录失败
        //Todo 因为要测试暂时先把这个登录失败去掉,等正式上线时打开
        log.error("wx auth failed, errCode is [{}], errMsg is [{}]",
wxLoginResult.getErrcode(), wxLoginResult.getErrmsg());
        /*throw new BasicException(
        ExceptionEnum.WX_AUTH_FAILED.customMessage(
                "wx auth failed, errCode is [%s], errMsg is [%s]",
                wxLoginResult.getErrcode(), wxLoginResult.getErrmsg()));*/
        wxLoginResult.setOpenid("123456789");
        wxLoginResult.setSession_key("123456789");
        wxLoginResult.setAccess_token("123456789");
        wxLoginResult.setUnionid("test");
    }
    //登录成功
    ...
    WxAppletAuthenticationToken wxAppletAuthenticationToken =
            new WxAppletAuthenticationToken(wxLoginResult.getOpenid(),
                    wxLoginResult.getSession_key(),
                    wxLoginResult.getAccess_token(),
                    wxLoginResult.getUnionid(),
                    wxLoginResult.getSource(),
                    userinfo.getNickname(), userinfo.getHeadimgurl(),
                    userinfo.getSex());
    wxAppletAuthenticationToken.setIp(ipAddress);
    return this.getAuthenticationManager().authenticate
(wxAppletAuthenticationToken);
```

(7) 修改Spring Security配置

以上内容创建完毕之后,需要在Spring Security中进行设置才能发挥作用。设置时,自定义WebSecurityConfig类,继承自WebSecurityConfigurerAdapter。重写configure方法,除了基本设置外,单独设置与JWT相关的内容。

```java
@Override
protected void configure(HttpSecurity http) throws Exception {
    http.csrf()
            .disable()
            .sessionManagement()
            // 不创建Session,使用JWT来管理用户的登录状态
            .sessionCreationPolicy(SessionCreationPolicy.STATELESS)
            .and()
            .authorizeRequests()
            // /error 异常端点不需要用户认证
            // 隐私协议不需要认证
            .antMatchers("/error/**","/vaccine/privacy/**","/JA82ybBt9x.txt",
"/vaccine/noauth/**").permitAll()
            // 其余的全部需要用户认证
```

```
            .anyRequest().authenticated()
            .and()
            .exceptionHandling()
            .authenticationEntryPoint(new CustomAuthenticationEntryPoint())
            .accessDeniedHandler(new CustomAccessDeniedHandler());
        // 使用WxAppletAuthenticationFilter替换默认的认证过滤器
UsernamePasswordAuthenticationFilter
        http.addFilterAt(wxAppletAuthenticationFilter(),
UsernamePasswordAuthenticationFilter.class)
        // 在WxAppletAuthenticationFilter前面添加用于验证JWT、识别用户是否登录的过滤器
            .addFilterBefore(jwtAuthenticationTokenFilter(),
WxAppletAuthenticationFilter.class);
    }
```

这里使用了无状态的Session，即使用JWT来管理用户登录状态。

```
.sessionCreationPolicy(SessionCreationPolicy.STATELESS)
```

所以要在登录过滤器之前引入JWT相关的过滤器。

```
.addFilterBefore(jwtAuthenticationTokenFilter(),
WxAppletAuthenticationFilter.class)
```

另外，由于微信小程序用户登录的特殊性，这里需要替换原来的用户名密码过滤器。

```
http.addFilterAt(wxAppletAuthenticationFilter(),
UsernamePasswordAuthenticationFilter.class)
```

以上声明的自定义过滤器和处理器需要在WebSecurityConfig中注入，否则无法正常工作。

```
//认证管理器
@Autowired
private WxAppletAuthenticationManager wxAppletAuthenticationManager;
//微信登录过滤器
@Bean
public WxAppletAuthenticationFilter wxAppletAuthenticationFilter(){
    WxAppletAuthenticationFilter wxAppletAuthenticationFilter = new
WxAppletAuthenticationFilter("/login");
    wxAppletAuthenticationFilter.setAuthenticationManager
(wxAppletAuthenticationManager);
    wxAppletAuthenticationFilter.setAuthenticationSuccessHandler
(customAuthenticationSuccessHandler());
    return wxAppletAuthenticationFilter;
}
//认证成功处理器
@Bean
public CustomAuthenticationSuccessHandler customAuthenticationSuccessHandler(){
    return new CustomAuthenticationSuccessHandler();
}
//JWT验证用户是否登录的过滤器
@Bean
public JwtAuthenticationTokenFilter jwtAuthenticationTokenFilter() {
    return new JwtAuthenticationTokenFilter();
}
```

经过以上配置之后，Spring Boot就成功集成了JWT，并同时与Spring Security进行了结合。

6.2 项目前期准备

6.2.1 项目需求说明

整个项目包含两部分内容：小程序端和小程序后台服务端。

小程序后台服务端，主要用来实现微信小程序用户登录，并能够与后台进行数据交互。

登录、数据提交以及查询数据时，需要小程序端来操作。

本章项目涉及的功能比较简单，重点在于MyBatis-Plus与JWT框架的使用，以及小程序与后台服务进行数据交互的实现。读者可以以当前项目为基础，根据业务场景自行扩展。

6.2.2 系统功能设计

根据以上项目需求，大致得出整个项目的结构，如图6.14所示。

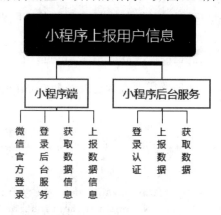

图6.14　小程序上报信息结构图

（1）小程序端

微信官方登录：调用微信官方登录接口，完成第一步登录，获取code。该code是后台服务与微信官方接口交互的凭证。

登录后台服务：需要携带code参数向后台发送登录请求。

获取数据信息：获取当前用户已提交的信息。

上报数据信息：提交当前用户相关的一些信息。

（2）小程序后台服务

登录认证：后台接收到客户端的登录请求后，根据获取的code向微信官方接口发送请求，验证用户信息，成功后再登录后台服务。

上报数据：接收到客户端提交的数据，保存。

获取数据：查询出当前用户提交的信息，返回到客户端。

6.2.3 系统数据库设计

这里的数据库主要涉及的是后台服务端，小程序客户端不涉及这方面的设计。

整个数据库包含两大部分，一部分是用户账号相关的数据表，另一部分是上报信息相关的数据表。

用户账号相关的数据表有：用户账号表（client_user）、角色表（client_role）、用户角色关联表（client_user_role）、权限表（client_permission）、角色权限关联表（client_role_permission）。

上报数据相关的数据表有：用户信息存储表（client_patient_info）、系统地址库表（sys_address）、数据字典相关表（sys_dict、sys_dict_type）。

1. 用户账号表client_user

该表用来存储登录当前系统的微信用户信息，包括微信openid、unionid、session_key、昵称、微信手机号、最后登录时间和IP、账号的状态等。

```
CREATE TABLE 'client_user' (
  'id' bigint(0) NOT NULL COMMENT 'id',
  'revision' int(0) NULL DEFAULT NULL COMMENT '乐观锁',
  'created_by' bigint(0) NULL DEFAULT NULL COMMENT '创建人',
  'created_time' datetime(0) NULL DEFAULT NULL COMMENT '创建时间',
  'updated_by' bigint(0) NULL DEFAULT NULL COMMENT '更新人',
  'updated_time' datetime(0) NULL DEFAULT NULL COMMENT '更新时间',
  'username' varchar(255) CHARACTER SET utf8 COLLATE utf8_general_ci NULL DEFAULT NULL COMMENT '姓名',
  'openid' varchar(255) CHARACTER SET utf8 COLLATE utf8_general_ci NULL DEFAULT NULL COMMENT '微信openid',
  'unionid' varchar(255) CHARACTER SET utf8 COLLATE utf8_general_ci NULL DEFAULT NULL COMMENT '微信开放平台唯一id',
  'session_key' varchar(255) CHARACTER SET utf8 COLLATE utf8_general_ci NULL DEFAULT NULL COMMENT '微信sessionkey',
  'app_openid' varchar(255) CHARACTER SET utf8 COLLATE utf8_general_ci NULL DEFAULT NULL COMMENT '微信开放平台app_openid',
  'avatar_url' varchar(255) CHARACTER SET utf8 COLLATE utf8_general_ci NULL DEFAULT NULL COMMENT '微信头像地址',
  'nick_name' varchar(255) CHARACTER SET utf8 COLLATE utf8_general_ci NULL DEFAULT NULL COMMENT '微信昵称',
  'phone_number' varchar(255) CHARACTER SET utf8 COLLATE utf8_general_ci NULL DEFAULT NULL COMMENT '微信手机号',
  'last_login_ip' varchar(255) CHARACTER SET utf8 COLLATE utf8_general_ci NULL DEFAULT NULL COMMENT '最后登录IP',
  'last_login_time' datetime(0) NULL DEFAULT NULL COMMENT '最后登录时间',
  'is_deleted' int(0) NOT NULL COMMENT '是否删除',
  'user_type' int(0) NULL DEFAULT NULL COMMENT '用户角色',
  'birthday' datetime(0) NULL DEFAULT NULL COMMENT '出生日期',
  'gender' int(0) NULL DEFAULT NULL COMMENT '性别',
  'password' varchar(255) CHARACTER SET utf8 COLLATE utf8_general_ci NULL DEFAULT NULL COMMENT '密码',
```

```
    'account' varchar(255) CHARACTER SET utf8 COLLATE utf8_general_ci NULL DEFAULT
NULL COMMENT '账号',
    'status' int(0) NULL DEFAULT NULL COMMENT '状态',
    'authed' int(0) NULL DEFAULT NULL COMMENT '授权',
    'authed_time' datetime(0) NULL DEFAULT NULL COMMENT '授权登录时间',
    'access_token' varchar(255) CHARACTER SET utf8 COLLATE utf8_general_ci NULL
DEFAULT NULL,
    PRIMARY KEY ('id') USING BTREE
) ENGINE = InnoDB CHARACTER SET = utf8 COLLATE = utf8_general_ci COMMENT = '客
户端用户表;客户端用户表' ROW_FORMAT = Dynamic;
```

2. 角色表client_role

角色表用来存储当前系统的角色信息，包括角色名称、创建和更新信息等。

```
CREATE TABLE 'client_user' (
    'id' bigint(0) NOT NULL COMMENT 'id',
    'revision' int(0) NULL DEFAULT NULL COMMENT '乐观锁',
    'created_by' bigint(0) NULL DEFAULT NULL COMMENT '创建人',
    'created_time' datetime(0) NULL DEFAULT NULL COMMENT '创建时间',
    'updated_by' bigint(0) NULL DEFAULT NULL COMMENT '更新人',
    'updated_time' datetime(0) NULL DEFAULT NULL COMMENT '更新时间',
    'username' varchar(255) CHARACTER SET utf8 COLLATE utf8_general_ci NULL DEFAULT
NULL COMMENT '姓名',
    'openid' varchar(255) CHARACTER SET utf8 COLLATE utf8_general_ci NULL DEFAULT
NULL COMMENT '微信openid',
    'unionid' varchar(255) CHARACTER SET utf8 COLLATE utf8_general_ci NULL DEFAULT
NULL COMMENT '微信开放平台唯一id',
    'session_key' varchar(255) CHARACTER SET utf8 COLLATE utf8_general_ci NULL
DEFAULT NULL COMMENT '微信sessionkey',
    'app_openid' varchar(255) CHARACTER SET utf8 COLLATE utf8_general_ci NULL DEFAULT
NULL COMMENT '微信开放平台app_openid',
    'avatar_url' varchar(255) CHARACTER SET utf8 COLLATE utf8_general_ci NULL DEFAULT
NULL COMMENT '微信头像地址',
    'nick_name' varchar(255) CHARACTER SET utf8 COLLATE utf8_general_ci NULL
NULL COMMENT '微信昵称',
    'phone_number' varchar(255) CHARACTER SET utf8 COLLATE utf8_general_ci NULL
DEFAULT NULL COMMENT '微信手机号',
    'last_login_ip' varchar(255) CHARACTER SET utf8 COLLATE utf8_general_ci NULL
DEFAULT NULL COMMENT '最后登录IP',
    'last_login_time' datetime(0) NULL DEFAULT NULL COMMENT '最后登录时间',
    'is_deleted' int(0) NOT NULL COMMENT '是否删除',
    'user_type' int(0) NULL DEFAULT NULL COMMENT '用户角色',
    'birthday' datetime(0) NULL DEFAULT NULL COMMENT '出生日期',
    'gender' int(0) NULL DEFAULT NULL COMMENT '性别',
    'password' varchar(255) CHARACTER SET utf8 COLLATE utf8_general_ci NULL DEFAULT
NULL COMMENT '密码',
    'account' varchar(255) CHARACTER SET utf8 COLLATE utf8_general_ci NULL DEFAULT
NULL COMMENT '账号',
    'status' int(0) NULL DEFAULT NULL COMMENT '状态',
    'authed' int(0) NULL DEFAULT NULL COMMENT '授权',
    'authed_time' datetime(0) NULL DEFAULT NULL COMMENT '授权登录时间',
```

```
  'access_token' varchar(255) CHARACTER SET utf8 COLLATE utf8_general_ci NULL
DEFAULT NULL,
  PRIMARY KEY ('id') USING BTREE
) ENGINE = InnoDB CHARACTER SET = utf8 COLLATE = utf8_general_ci COMMENT = '客
户端用户表;客户端用户表' ROW_FORMAT = Dynamic;
```

3. 用户角色关联表client_user_role

这个表用来存储系统登录用户与角色之间的关联关系。

```
CREATE TABLE 'client_user_role' (
  'id' bigint(0) NOT NULL AUTO_INCREMENT,
  'client_user_id' bigint(0) NULL DEFAULT NULL,
  'role_id' bigint(0) NULL DEFAULT NULL,
  'is_deleted' tinyint(0) NULL DEFAULT NULL COMMENT '删除状态, 0：未删除, 1：已删除',
  'created_time' datetime(0) NOT NULL DEFAULT CURRENT_TIMESTAMP,
  'updated_time' datetime(0) NOT NULL DEFAULT CURRENT_TIMESTAMP,
  'created_by' bigint(0) NULL DEFAULT NULL,
  'updated_by' bigint(0) NULL DEFAULT NULL,
  'revision' int(0) NULL DEFAULT NULL COMMENT '乐观锁',
  PRIMARY KEY ('id') USING BTREE
) ENGINE = InnoDB AUTO_INCREMENT = 1511636011313274881 CHARACTER SET = utf8mb4
COLLATE = utf8mb4_general_ci ROW_FORMAT = Dynamic;
```

4. 权限表client_permission

权限表用来存储当前系统的权限，包括权限名称、创建和更新信息等。

```
CREATE TABLE 'client_permission' (
  'id' bigint(0) NOT NULL AUTO_INCREMENT,
  'permission' varchar(32) CHARACTER SET utf8mb4 COLLATE utf8mb4_general_ci NULL
DEFAULT NULL COMMENT '资源权限，如：user:list,user:create',
  'is_deleted' tinyint(0) NULL DEFAULT NULL COMMENT '删除状态, 0：未删除, 1：已删除',
  'created_time' datetime(0) NOT NULL DEFAULT CURRENT_TIMESTAMP COMMENT '创建时间',
  'updated_time' datetime(0) NOT NULL DEFAULT CURRENT_TIMESTAMP COMMENT '更新时间',
  'created_by' bigint(0) NULL DEFAULT NULL COMMENT '创建者id',
  'updated_by' bigint(0) NULL DEFAULT NULL COMMENT '更新者id',
  'revision' int(0) NULL DEFAULT NULL COMMENT '乐观锁',
  PRIMARY KEY ('id') USING BTREE
) ENGINE = InnoDB AUTO_INCREMENT = 45 CHARACTER SET = utf8mb4 COLLATE =
utf8mb4_general_ci ROW_FORMAT = Dynamic;
```

5. 角色权限关联表client_role_permission

该表用来存储角色和权限之间的关联关系。

```
CREATE TABLE 'client_role_permission' (
  'id' bigint(0) NOT NULL AUTO_INCREMENT,
  'role_id' bigint(0) NULL DEFAULT NULL COMMENT '角色id',
  'permission_id' bigint(0) NULL DEFAULT NULL COMMENT '权限id',
  'is_deleted' tinyint(0) NULL DEFAULT NULL COMMENT '删除状态, 0：未删除, 1：已删除',
  'created_time' datetime(0) NOT NULL DEFAULT CURRENT_TIMESTAMP COMMENT '创建时间',
  'updated_time' datetime(0) NOT NULL DEFAULT CURRENT_TIMESTAMP COMMENT '更新时间',
  'created_by' bigint(0) NULL DEFAULT NULL COMMENT '创建者id',
```

```
  'updated_by' bigint(0) NULL DEFAULT NULL COMMENT '更新者id',
  'revision' int(0) NULL DEFAULT NULL COMMENT '乐观锁',
  PRIMARY KEY ('id') USING BTREE
) ENGINE = InnoDB AUTO_INCREMENT = 34 CHARACTER SET = utf8mb4 COLLATE = utf8mb4_general_ci ROW_FORMAT = Dynamic;
```

6. 用户信息存储表client_patient_info

这个表用来存储用户账号提交的个人信息，包括真实姓名、联系电话、证件信息、出生日期、性别、居住地、身高、体重、其他联系信息等。

```
CREATE TABLE 'client_patient_info' (
  'id' bigint(0) NOT NULL COMMENT 'id',
  'is_deleted' int(0) NULL DEFAULT NULL,
  'revision' int(0) NULL DEFAULT NULL COMMENT '乐观锁',
  'created_by' bigint(0) NULL DEFAULT NULL COMMENT '创建人',
  'created_time' datetime(0) NULL DEFAULT NULL COMMENT '创建时间',
  'updated_by' bigint(0) NULL DEFAULT NULL COMMENT '更新人',
  'updated_time' datetime(0) NULL DEFAULT NULL COMMENT '更新时间',
  'client_user_id' bigint(0) NULL DEFAULT NULL COMMENT 'user id',
  'real_name' varchar(255) CHARACTER SET utf8 COLLATE utf8_general_ci NULL DEFAULT NULL COMMENT '真实姓名',
  'phone' varchar(255) CHARACTER SET utf8 COLLATE utf8_general_ci NULL DEFAULT NULL COMMENT '联系电话',
  'card_type' bigint(0) NULL DEFAULT NULL COMMENT '证件类型',
  'card_num' varchar(255) CHARACTER SET utf8 COLLATE utf8_general_ci NULL DEFAULT NULL COMMENT '证件号',
  'birthday' datetime(0) NULL DEFAULT NULL COMMENT '出生日期',
  'gender' int(0) NULL DEFAULT NULL COMMENT '性别;1男2女',
  'province_id' bigint(0) NULL DEFAULT NULL COMMENT '居住地省份',
  'city_id' bigint(0) NULL DEFAULT NULL COMMENT '居住地城市',
  'district_id' bigint(0) NULL DEFAULT NULL COMMENT '居住地区域id',
  'detail_address' varchar(255) CHARACTER SET utf8 COLLATE utf8_general_ci NULL DEFAULT NULL COMMENT '详细地址',
  'residence_province_id' bigint(0) NULL DEFAULT NULL COMMENT '户口所在地省份',
  'residence_city_id' bigint(0) NULL DEFAULT NULL COMMENT '户口所在地城市',
  'residence_district_id' bigint(0) NULL DEFAULT NULL COMMENT '户口所在地区、乡镇等',
  'body_height' decimal(24, 2) NULL DEFAULT NULL COMMENT '身高',
  'body_weight' decimal(24, 2) NULL DEFAULT NULL COMMENT '体重',
  'occupation_id' bigint(0) NULL DEFAULT NULL COMMENT '职业或专业',
  'email' varchar(255) CHARACTER SET utf8 COLLATE utf8_general_ci NULL DEFAULT NULL COMMENT '电子邮箱',
  'wechat_account' varchar(255) CHARACTER SET utf8 COLLATE utf8_general_ci NULL DEFAULT NULL COMMENT '微信号',
  PRIMARY KEY ('id') USING BTREE
) ENGINE = InnoDB CHARACTER SET = utf8 COLLATE = utf8_general_ci COMMENT = '基本信息表' ROW_FORMAT = Dynamic;
```

7. 系统地址库表sys_address

全国地区地址库，分省、市、区三级。

```sql
CREATE TABLE 'sys_address' (
  'id' bigint(0) NULL DEFAULT NULL,
  'name' varchar(255) CHARACTER SET utf8 COLLATE utf8_general_ci NULL DEFAULT NULL COMMENT '名称',
  'fullname' varchar(255) CHARACTER SET utf8 COLLATE utf8_general_ci NULL DEFAULT NULL COMMENT '全称,即携带省、市或区名',
  'parent_id' bigint(0) NULL DEFAULT NULL COMMENT '上级地址id',
  'province_id' bigint(0) NULL DEFAULT NULL COMMENT '所属省份id',
  'level' varchar(255) CHARACTER SET utf8 COLLATE utf8_general_ci NULL DEFAULT NULL COMMENT '级别,1:省或直辖市,2:市,3:区或县',
  'tel_code' varchar(255) CHARACTER SET utf8 COLLATE utf8_general_ci NULL DEFAULT NULL COMMENT '电话区号'
) ENGINE = InnoDB CHARACTER SET = utf8 COLLATE = utf8_general_ci ROW_FORMAT = Dynamic;
```

8. 数据字典相关表sys_dict和sys_dict_type

sys_dict_type用来存储数据字典的类型。sys_dict用来存储数据字典具体的数据列表,例如职业列表。

```sql
CREATE TABLE 'sys_dict_type' (
  'id' bigint(0) NOT NULL COMMENT 'id',
  'revision' int(0) NULL DEFAULT NULL COMMENT '乐观锁',
  'created_by' bigint(0) NULL DEFAULT NULL COMMENT '创建人',
  'created_time' datetime(0) NULL DEFAULT NULL COMMENT '创建时间',
  'updated_by' bigint(0) NULL DEFAULT NULL COMMENT '更新人',
  'updated_time' datetime(0) NULL DEFAULT NULL COMMENT '更新时间',
  'dict_type_key' varchar(255) CHARACTER SET utf8 COLLATE utf8_general_ci NULL DEFAULT NULL COMMENT '数据字典分组的key',
  'is_deleted' int(0) NULL DEFAULT NULL COMMENT '是否删除',
  'fixed' int(0) NULL DEFAULT NULL COMMENT '是否为初始化固定的,固定的不可修改',
  'dict_type_name' varchar(255) CHARACTER SET utf8 COLLATE utf8_general_ci NULL DEFAULT NULL COMMENT '类型名称',
  PRIMARY KEY ('id') USING BTREE
) ENGINE = InnoDB CHARACTER SET = utf8 COLLATE = utf8_general_ci COMMENT = '数据字典分组表' ROW_FORMAT = Dynamic;
CREATE TABLE 'sys_dict' (
  'id' bigint(0) NOT NULL COMMENT 'id',
  'revision' int(0) NULL DEFAULT NULL COMMENT '乐观锁',
  'created_by' bigint(0) NULL DEFAULT NULL COMMENT '创建人',
  'created_time' datetime(0) NULL DEFAULT NULL COMMENT '创建时间',
  'updated_by' bigint(0) NULL DEFAULT NULL COMMENT '更新人',
  'updated_time' datetime(0) NULL DEFAULT NULL COMMENT '更新时间',
  'dict_type_id' bigint(0) NULL DEFAULT NULL COMMENT '数据分类,职业类别,基因突变类型',
  'order_num' int(0) NULL DEFAULT NULL COMMENT '排序号',
  'dict_name' varchar(255) CHARACTER SET utf8 COLLATE utf8_general_ci NULL DEFAULT NULL COMMENT '数据项名称',
  'is_deleted' int(0) NULL DEFAULT NULL COMMENT '是否删除',
  'data_value' varchar(255) CHARACTER SET utf8 COLLATE utf8_general_ci NULL DEFAULT NULL COMMENT '具体取值',
  PRIMARY KEY ('id') USING BTREE
) ENGINE = InnoDB CHARACTER SET = utf8 COLLATE = utf8_general_ci COMMENT = '数据字典表' ROW_FORMAT = Dynamic;
```

6.2.4 系统文件说明

1. 后台服务系统文件说明

本项目中后台服务系统文件的组织结构如图6.15所示。

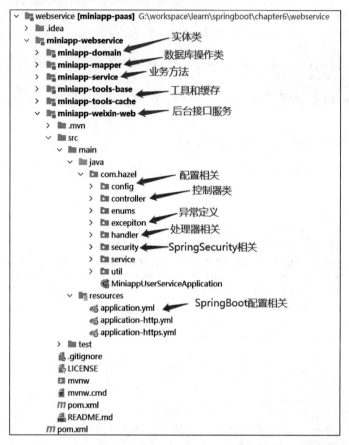

图6.15 后台服务系统文件组织结构

整个后台服务分为6部分，各部分的说明如下：

- miniapp-domain：实体相关类，例如与数据库表对应的Entity实体、查询结果映射的VO对象、数据传递的DTO对象、查询条件相关的QO对象等，均在此模块中。
- miniapp-mapper：数据库操作类，与数据库进行数据交互，进行数据的增、删、改、查。
- miniapp-service：业务相关类，以事务层面开启业务处理，调用mapper层提供的接口。
- miniapp-tools-base：公用工具类，通用返回封装的对象、错误码定义等。
- miniapp-tools-cache：缓存相关的工具类，支持各种常见类型的数据缓存。
- miniapp-weixin-web：整个项目的核心工程，小程序后台的接口服务模块，也是配置Spring Boot、Spring Security、JWT、MyBatis-Plus的模块。

2. 小程序端项目系统文件说明

除了后台服务端，我们还有小程序客户端，它的文件组织结构如图6.16所示。
各部分说明如下：

- common：通用部分，例如地图插件、图表插件等。
- components：组件相关，例如选择器组件等。
- core：请求接口的域名和地址配置，各种请求方式的通用方法封装。
- node_modules：Node.js库，安装后会自动出现。
- pages：小程序端页面，有首页、我的、完善信息等页面。
- static：静态资源，例如图片、图标、CSS样式文件等。
- uni-modules：HBuilderX中uni-app的相关插件，例如日期插件、常用图标等。

图6.16 小程序端系统文件组织结构

6.3 项目前端设计

6.3.1 首页

首页展示一些小程序中常用的模块，这里给出的模块是问答、百科和视频模块，读者可以根据业务自行扩展模块。

整个首页分为两部分，上半部分是问答和百科，下半部分是教程和视频，如图6.17所示。每一个模块直接使用view布局即可。

```
<view class="dmd-index-top-block" @click="introduction">
    <image src="../../static/images/baike.png"/>
    <view class="dmd-index-top-title">看百科</view>
    <view class="dmd-index-top-hint">科普</view>
</view>
```

在CSS部分定义好样式。

```
.dmd-index-main-block-line{
    display: flex;
    justify-content: space-around;
    flex-direction: row;
    height:200rpx;
    .dmd-index-main-block {
        height: 200rpx;
        width: 31%;
        text-align: center;
    }
}
```

图6.17 首页

.dmd-index-main-block-line代表一行的样式，如果想定义这一行内每一个图标的样式，就可以直接在它的内部定义，例如.dmd-index-main-block。

6.3.2 我的

从"我的"页面可以跳转至微信一键登录，也可以跳转至当前用户的信息完善页面，"我的"页面如图6.18所示。

6.3.3 微信一键登录

在微信一键登录页面上点击"微信登录授权"按钮触发登录请求，如图6.19所示。

图6.18 "我的"页面　　　　　图6.19 微信授权登录

用户点击"微信登录授权"按钮后向微信官方后台发送登录请求，获取code，得到code之后，携带该code登录后台系统，向后台发送登录请求，得到授权token，然后将该授权token存储在小程序缓存中，用于后续接口的调用。

```
// 获取用户code码
login() {
    uni.login({
        onlyAuthorize:true,
        provider:'weixin',
        success: (e) => {
            if (e.errMsg == 'login:ok') {
                this.code = e.code
                this.getOpenid()
            }
```

```
        }
    })
}
//向后台发送登录请求，得到授权信息
async getOpenid() {
    let params = {code: this.code} //code参数为上一步从微信官方获取的code
    const {
        data
    } = await this.$request({
        url: '/login',//后台登录接口API地址
        method: 'POST',
        data: params
    })
    if (data) {
        this.Authorization = data.Authorization //登录成功后的授权信息，即JWT token
        uni.setStorageSync('account', data.account)  //缓存该微信用户在后台对应的账号
        uni.setStorageSync('Authorization', data.Authorization) //缓存授权信息
    } else {
        ...
    }
}
```

6.3.4 完善信息

"完善信息"页面用于提交当前用户的相关信息，如果已提交用户信息，则会展示当前用户的信息用于修改，如图6.20、图6.21所示。

图6.20 "完善信息"页面

图6.21 修改信息

这里用到了两个插件：日期选择插件和地址选择插件。

```
<view class="item">
    <view class="item-top">
        <text class="star"></text>
        <text class="title">出生日期</text>
    </view>
    <view class="item-form">
        <input class="item-form-input" type="text" placeholder="请选择出生日期" placeholder-class="placeholder"
            v-model="formData.birthday" disabled/>
        <text class="choose" @click="birthdayVisibleClick">选择</text> //选择按钮
    </view>
</view>
<!--出生日期插件-->
<w-picker :visible.sync="birthdayVisible" mode="date" startYear="1930" endYear="2050"
        :value="birthdayInitValue" fields="day" @confirm="onConfirm($event, 'birthday')"
        :disabled-after="true" ref="birthday"> //picker日期插件
</w-picker>
<view class="item-form">
<uni-data-picker :localdata="areas" preload v-model="formData.districtId":map="{text:'name',value:'id'}"
    popup-title="请选择省市区" @change="addressChange" @nodeclick="addressNodeClick" ref="addressPicker"
            v-slot:default="{data, error, options}"> //data-picker数据选择插件
<view v-if="error" class="error">
<text>{{error}}</text>
</view>
<view v-else-if="data.length" class="selected">
<text v-for="(item,index) in data" :key="index">{{item.text + " "}}</text>//遍历省市区数据
</view>
<view v-else>
<text class="address-choose">请选择省市区</text>//提示语
</view>
</uni-data-picker>
<text class="choose" @click="addressPickerShow">选择</text>//选择按钮
</view>
```

6.3.5　底部导航栏

底部导航栏包含两个模块，"首页"和"我的"，如图6.22所示。

图6.22　小程序底部导航

在pages.json页面设置tabBar。

```
    "tabBar": {
        "color": "#8a8a8a",
        "selectedColor": "#efb336",
        "borderStyle": "black",
        "backgroundColor": "#ffffff",
        "list": [{
            "pagePath": "pages/index/index",
            "iconPath": "static/home.png",
            "selectedIconPath": "static/home_active.png",
            "text": "首页"
        },
        {
            "pagePath": "pages/my/index",
            "iconPath": "static/my.png",
            "selectedIconPath": "static/my_active.png",
            "text": "我的"
        }]
    }
```

6.4 项目后端实现

6.4.1 JWT登录认证

JWT登录认证用来验证当前用户是否已登录。已登录用户可以继续向下执行，其他不符合认证的请求都将被拦截返回。

认证过程中，导致JWT失效的一般有三种情况：

- 非法token：无法解析出正确的值。
- token过期：token已超过签发的有效期。
- 签名无效：签名不匹配，无法正确解析。

```
protected void doFilterInternal(HttpServletRequest request, HttpServletResponse
response, FilterChain filterChain) throws ServletException, IOException {
    log.debug("processing authentication for [{}]", request.getRequestURI());
    String token = request.getHeader(ConstantEnum.AUTHORIZATION.getValue());//
获取认证的token令牌
    String unionid = null;//用户唯一识别码，没有unionid的可以用openid
    if (token != null) {//token存在
        try {
            unionid = jwtTokenUtils.getUsernameFromToken(token);//解析出唯一识别码
        } catch (IllegalArgumentException e) {//token格式不正确，无法解析
            log.error("an error occurred during getting username from token", e);
            throw new BasicException(ExceptionEnum.JWT_EXCEPTION.customMessage
("an error occurred during getting username from token , token is [%s]", token));
        } catch (ExpiredJwtException e) {//token已过期
            log.warn("the token is expired and not valid anymore", e);
```

```java
            throw new BasicException(ExceptionEnum.JWT_EXCEPTION.customMessage
("the token is expired and not valid anymore, token is [%s]", token));
        }catch (SignatureException e) {//签名不匹配
            log.warn("JWT signature does not match locally computed signature", e);
            throw new BasicException(ExceptionEnum.JWT_EXCEPTION.customMessage
("JWT signature does not match locally computed signature, token is [%s]", token));
        }
    }else {
        log.warn("couldn't find token string");//没有token
    }
    //已登录
    if (unionid != null && SecurityContextHolder.getContext().getAuthentication()
== null) {
        log.debug("security context was null, so authorizing user");
        ClientUser account = authService.findAccountByUnionid(unionid);//查找用户
信息
        //获取用户权限
        List<ClientPermission> permissions = authService.acquirePermission
(account.getId());
        List<SimpleGrantedAuthority> authorities =
permissions.stream().map(permission -> new
SimpleGrantedAuthority(permission.getPermission())).collect(Collectors.toList());
        log.info("authorized user [{}], setting security context", unionid);
        //将用户与权限信息放入认证容器
        SecurityContextHolder.getContext().setAuthentication(
            new WxAppletAuthenticationToken(unionid,authorities));
    }
    //继续向下执行过滤器
    filterChain.doFilter(request, response);
}
```

6.4.2 登录与注册

由于本项目场景的特殊性，登录不再以普通的账号密码形式登录，而是携带微信官方提供的code进行认证。

```java
//发送请求到微信平台接口，获取登录结果
WxLoginResultDTO wxLoginResult = restTemplate.getForObject(url,
WxLoginResultDTO.class);
wxLoginResult.setSource(source);
if(wxLoginResult.getErrcode() != null
&& !wxLoginResult.getErrcode().equals(0)){//登录失败
    log.error("wx auth failed, errCode is [{}], errMsg is [{}]",
wxLoginResult.getErrcode(), wxLoginResult.getErrmsg());
}
```

后台服务接收到小程序端的请求之后，首先调用过滤器，请求微信平台官方接口，获取登录结果，传递的参数为小程序端发送过来的code。如果登录结果是失败，则返回授权失败，登录失败。如果登录结果是成功，那么继续向下进行本平台的登录流程，也就是进行Spring Security的认证。

```java
        WxAppletAuthenticationToken wxAppletAuthenticationToken = null;
        log.info("account is {}",authentication);
        if (authentication instanceof WxAppletAuthenticationToken) {
            wxAppletAuthenticationToken = (WxAppletAuthenticationToken) authentication;
        }
        String openid = wxAppletAuthenticationToken.getOpenid();  //微信平台openid,在小程
序中唯一
        String sessionKey = wxAppletAuthenticationToken.getSessionKey();    //登录后的
sessionkey
        String accessToken = wxAppletAuthenticationToken.getAccessToken(); //登录后的
accessToken
        String source = wxAppletAuthenticationToken.getSource();//登录用户来源,小程序还是
app,目前只有小程序可以先不管这个
        String headimgurl = wxAppletAuthenticationToken.getHeadimgurl();      //头像地址
        Integer sex = wxAppletAuthenticationToken.getSex();                   //性别
        String nickname = wxAppletAuthenticationToken.getNickname();          //昵称
        // 查找本地后台中的账号
        ClientUser account = null;
        if(StringUtils.isNotBlank(wxAppletAuthenticationToken.getUnionid())){    //没有
unionid,就以openid为准
            account = authService.findAccountByUnionid
(wxAppletAuthenticationToken.getUnionid());
        }
        if(account == null){
           account = authService.findAccount
(wxAppletAuthenticationToken.getOpenid(),source);
        }
        //执行注册逻辑
        if (account == null) {
            log.debug("account not exist, began to register. openid is [{}]",
wxAppletAuthenticationToken.getOpenid());
            //签名校验
            Digester digester = new Digester(DigestAlgorithm.SHA1);
            String data = wxAppletAuthenticationToken.getRawData() +
wxAppletAuthenticationToken.getSessionKey();
            String signature = digester.digestHex(data);
            if (!wxAppletAuthenticationToken.getSignature().equals(signature)) {
                log.error("signature is invalid, [{}] vs [{}]", signature,
wxAppletAuthenticationToken.getSignature());
                throw new BasicException(ExceptionEnum.SIGN_INVALID.customMessage
("signature is invalid, [%s] vs [%s]", signature,
wxAppletAuthenticationToken.getSignature()));
            }
            //获取用户信息
            account = new ClientUser();
            if(StringUtils.isNotBlank(source) && source.equals("app")){//App登录过来的用户
                account.setAppOpenid(openid);
                account.setAccessToken(accessToken);
            }else{//小程序登录的用户
                account.setOpenid(openid);
                account.setSessionKey(sessionKey);
```

```
    }
    //以下为设置账号的基本信息
    account.setIsDeleted(0);
    account.setCreatedTime(DateUtil.date().toTimestamp());
    account.setLastLoginTime(DateUtil.date().toTimestamp());//更新最后登录时间
    account.setLastLoginIp(wxAppletAuthenticationToken.getIp());//更新最后登录IP
    account.setStatus(0);//更新状态
    account.setUnionid(wxAppletAuthenticationToken.getUnionid());//设置unionid,
如果有的话
    account.setAvatarUrl(headimgurl);//设置头像地址
    account.setGender(sex);//设置性别
    account.setNickName(nickname);//设置昵称
    log.info("account is [{}]", account);
    //注册用户
    account = authService.register(account);
    //获取权限
    List<ClientPermission> permissions = authService.acquirePermission
(account.getId());
    List<SimpleGrantedAuthority> authorities =
permissions.stream().map(permission -> new SimpleGrantedAuthority
(permission.getPermission())).collect(Collectors.toList());
    return new WxAppletAuthenticationToken(account.getOpenid(),
account.getSessionKey(),
        account.getAccessToken(),
        account.getUnionid(), wxAppletAuthenticationToken.getSource(),
        authorities);
}
```

这里的认证采用的是自定义WxAppletAuthenticationManager替换原本的账号密码认证管理器。在主要的认证方法中，首先拿到上一步过滤器返回的微信登录对象，根据返回的openid去数据库验证当前用户中是否存在该用户。如果存在，则首先更新用户的一些信息，例如最后登录时间和IP，然后将该用户的具体信息与生成的JWT认证token一同返回给小程序端。如果不存在，首先创建用户信息，存储微信官方返回的信息内容，然后生成新的用户信息并和JWT认证token一同返回给小程序端。

6.4.3 获取信息

获取信息时，首先获取当前登录用户，然后使用MyBatis-Plus提供的条件构造器构造查询条件，从数据库中查询用户信息。

```
//根据openid获取用户信息
ClientUser uploadUser = getLoginAccount();
if(uploadUser.getId() == null){
    return CommonResponse.error("当前用户不存在");
}
//使用MyBatis-Plus的条件构造器
LambdaQueryWrapper<ClientPatientInfo> wrapper = Wrappers.lambdaQuery();
wrapper.eq(ClientPatientInfo::getClientUserId,uploadUser.getId());
ClientPatientInfo info = Optional.ofNullable(clientInfoService.getOne(wrapper)
```

```
            .orElse(new ClientPatientInfo());
    return new CommonObjectResponse<>(info);
```

条件构造器的运算规则可以参考前面6.1.1节中关于条件构造器的介绍。这里使用的是等于eq，即用户id等于当前登录用户的id就是要查询的用户信息。

另外，这里还用到了Java 8版本之后出现的Lambda表达式功能ClientPatientInfo::getClientUserId，更加方便编写代码。

6.4.4 完善或修改信息

用户完成登录后，可以完善自己的信息，也可以修改自己的信息。下面代码是一个数据交互功能的简单演示，如果需要其他复杂业务，可以在此基础上扩展。后台接收到用户提交的信息后，需要单独处理的是省、市、区相关的信息。

```
//获取当前登录用户的信息
ClientUser uploadUser = getLoginAccount();
if(uploadUser.getId() == null){
    return CommonResponse.error("当前用户不存在");
}
//处理数据
//处理居住地省、市、区
Long districtId = patientInfo.getDistrictId();                      //居住区
SysAddress address = addressService.getById(districtId);            //居住区对应的对象
patientInfo.setProvinceId(address.getProvinceId());                 //居住地省ID
patientInfo.setCityId(address.getParentId());                       //居住地市ID
//处理户口所在地省、市、区
Long residenceDistrictId = patientInfo.getResidenceDistrictId();
SysAddress residenceAddress = addressService.getById(residenceDistrictId);
patientInfo.setResidenceProvinceId(residenceAddress.getProvinceId());
patientInfo.setResidenceCityId(residenceAddress.getParentId());
//获取当前用户已提交的信息
LambdaQueryWrapper<ClientPatientInfo> wrapper = Wrappers.lambdaQuery();
wrapper.eq(ClientPatientInfo::getClientUserId,uploadUser.getId());
patientInfo.setClientUserId(uploadUser.getId());
ClientPatientInfo info = clientInfoService.getOne(wrapper);
//未提交的,信息为null
if(info != null){
    //更新提交的信息
    BeanUtils.copyProperties(patientInfo,info,"id","isDeleted","createdBy",
"createdTime");
    updateBaseInfo(info, uploadUser.getId());
    clientInfoService.updateById(info);
}else{
    //未提交过信息的用户,完善用户信息
    buildBaseInfo(patientInfo, uploadUser.getId());
    clientInfoService.save(patientInfo);
}
```

这个方法集完善和修改于一体，因此不再另外定义修改的方法。

6.5 项目总结

本章实现了一个小程序上报信息的项目，后端使用的主要技术有Spring、Spring MVC、Spring Security、MyBatis-Plus和JWT框架，前端使用的技术主要有uni-app、Node.js和Vue，数据库使用的是MySQL。本章读者要掌握的重点是JWT框架和MyBatis-Plus的使用，以及如何通过Spring Boot整合这些框架并做相应配置、如何在实际开发过程中使用这些框架。学习时可同时参考各个框架官方提供的API手册，并且也要多加实践。

本章的重点技术：

- JWT框架：JWT的原理、优缺点、使用场景，以及使用Spring Boot集成Spring Security和JWT框架的方法。
- MyBatis-Plus数据库：MyBatis-Plus的特性以及功能，学会在项目中使用该框架。
- HBuilderX工具：工具的安装以及使用该工具搭建项目。
- 微信开发工具：使用微信开发工具打开HBuilderX编译过的代码，并简单调试功能。
- 小程序项目搭建所准备的内容：小程序官方账号、域名、证书等，仅作为学习使用不发布的情况下，可以不注册官方账号，只使用测试账号。

第 7 章

模拟聊天室

在第6章中，我们介绍了使用Spring Boot开发小程序服务接口的项目。本章将使用Spring Boot实现一个模拟聊天室，实现聊天用户登录、上线、下线、发送消息、群发消息等功能。模拟聊天室使用的主要技术是WebSocket。为了专注于WebSocket的学习，本章没有使用其他的新技术，全部都是沿用之前的技术。

本章主要涉及的知识点有：

- 如何使用Spring Boot集成WebSocket。
- 如何在Spring Boot中配置WebSocket。
- 前端页面与后台服务如何进行WebSocket消息交互。

7.1 项目技术选型

本章将实现一个简单的模拟聊天室，使用到的协议框架为WebSocket。Web框架为Spring，数据库采用MySQL，前端页面采用LayUI框架，页面模板直接使用HTML5。本项目的业务范围比较简单，主要讲解所用地框架技术，整个项目未采用前后端分离技术。

7.1.1 WebSocket

1. 概述

在Web开发领域，我们最常用的协议是HTTP，而HTTP协议有一个缺陷，通信只能由客户端发起，并不具备从服务端主动推送消息的功能，这也导致在浏览器端想要做到服务器主动推送的效果，只能用一些轮询和长轮询的方案来做，既耗时又无法达到高性能。

举例来说，物联网中某个监测实时温度的设备会不停地向服务器推送最新温度，前端页面如果想展示最新温度，只能是客户端向服务器发出请求，服务器返回查询结果，而如果想达到实时性的要求，必然要提高请求的频率，十秒一次乃至一秒一次，这就是轮询。如果这期间设备数据一直没有更新,势必造成资源浪费,而如果设备更新的频率大于轮询的频率,势必又会造成数据展示的延迟。

这种单向请求注定了如果服务器有连续的状态变化,客户端要获知相关信息就非常麻烦。这会导致一些问题:

- 服务端被迫维持来自每个客户端的大量不同的连接。
- 大量的轮询请求会造成高开销,比如会带上多余的header,造成了无用的数据传输。

HTTP协议本身是没有持久通信能力的,但是我们在实际的应用中又很需要这种能力,为了解决这些问题,WebSocket协议由此而生。HTTP协议和WebSocket协议都是基于TCP所做的封装。WebSocket协议于2011年被IETF定为标准RFC6455,并被RFC7936补充规范,而且在HTML5标准中增加了有关WebSocket协议的相关API,所以只要实现了HTML5标准的客户端,就可以与支持WebSocket协议的服务器进行全双工的持久通信了。目前WebSocket协议得到了所有主流浏览器的支持,同时它还兼容了HTTP协议,默认使用HTTP的80端口和443端口,同时使用HTTP header进行协议升级。

在WebSocket API中,浏览器和服务器只需要完成一次握手就直接可以创建持久性的连接,并进行双向数据传输。

WebSocket协议的其他特点包括:

- 建立在TCP协议之上,服务器端的实现比较容易。
- 与HTTP协议有着良好的兼容性。
- 数据格式比较轻量,性能开销小,通信高效。
- 可以发送文本,也可以发送二进制数据。
- 没有同源限制,客户端可以与任意服务器通信。
- 协议标识符是ws(如果加密,则为wss),服务器网址就是URL。
- 实时性更强。
- 有状态,开启链接之后可以不用每次都携带状态信息。

2. WebSocket协议原理

与HTTP协议一样,WebSocket协议也需要通过已建立的TCP连接来传输数据。具体实现上是先通过HTTP协议建立通道,即实现握手,然后在此基础上用真正的WebSocket协议进行通信。

首先来看握手协议。

```
Accept-Encoding: gzip, deflate
Accept-Language: zh-CN,zh;q=0.9
Cache-Control: no-cache
Connection: Upgrade
Host: xxx.xxx.xxx.xxx:8008
Origin: http://xxx.xxx.xxx.xxx:8008
Pragma: no-cache
Sec-WebSocket-Extensions: permessage-deflate; client_max_window_bits
Sec-WebSocket-Key: pQCUNY4AlNszv6mH7p1v7g==
```

```
Sec-WebSocket-Version: 13
Upgrade: websocket
User-Agent: Mozilla/5.0 (Windows NT 10.0; Win64; x64) AppleWebKit/537.36 (KHTML,
like Gecko) Chrome/102.0.0.0 Safari/537.36
```

协议参数说明：

- Connection：设置为Upgrade，表示为逐跳头部，用于服务器代理识别连接。
- Upgrade：设置为websocket，表示要将协议升级到WebSocket，而不是普通的HTTP请求。
- Sec-WebSocket-Key：这是一个Base64编码后的值，是浏览器随机生成的，用来验证WebSocket助理的真实性。
- Sec-WebSocket-Version：所使用的 WebSocket Draft（协议版本）。

服务器端接收到如上请求后，会返回如下内容。

```
Connection: upgrade
Date: Wed, 29 Jun 2022 07:40:55 GMT
Sec-WebSocket-Accept: +EyAq4Ha6euPTnkUFyvRpi2zJ+Q=
Sec-WebSocket-Extensions: permessage-deflate;client_max_window_bits=15
Server: nginx/1.21.5
Upgrade: websocket
```

以上内容表示，服务器端告诉客户端协议已经成功升级为WebSocket协议。客户端成功接收到消息后，与服务器开启全双工通信。

3. 使用场景

表7.1描述的是WebSocket协议常用的业务场景。除了以下这些场景外，还有很多其他的场景，这些场景基本都对数据有较高的实时性要求，这时候我们会用到WebSocket协议。本章项目选取的聊天室就是WebSocket即时通信的一个典型应用。

表7.1　WebSocket常用业务场景

业务场景	场景概述
弹幕	多个终端手机之间弹幕需要同步
在线教育	老师进行一对多的在线授课，在客户端内编写的笔记、大纲等信息需要实时推送至多个学生的客户端，这需要通过WebSocket协议来完成
股票价格	股票等价格变化迅速，可以通过WebSocket协议将变化后的价格实时推送至世界各地的客户端，方便交易员迅速做出交易判断
实况更新	体育实况消息以及在线实时新闻
视频会议和即时通信	视频聊天、视频会议、语音聊天、文字聊天等
实时位置	借用移动设备的GPS功能来实现基于位置的网络应用，例如地图导航、运动软件、户外考勤等
多玩家游戏	一些需要多个玩家共同操作的游戏，每个玩家都需要将自己的动态实时更新到别的玩家客户端上
在线协同编辑	多人在线实时编辑文档
物联网设备数据展示更新	实时监测设备，例如水位、温度等，当水位和温度发生变化时，需要实时推送到展示平台

4. Java中使用WebSocket

在Java中使用WebSocket有多种方式，一般常用的有以下几种：

- Javax包提供的WebSocket扩展：这种方式需要Tomcat的支持，所以需要导入相应的Tomcat依赖包。
- Spring实现的标准WebSocket：Spring 4.0以上版本支持WebSocket扩展，同时对浏览器也做了兼容。
- Spring实现的STOMP封装组件：STOMP消息功能在WebSocket基础上做了扩展，使用方式更加简便。
- Netty方式：在Netty组件中对常见协议进行了封装，其中就包括WebSocket。这种方式目前是许多物联网项目比较常用的方式。

（1）Javax包提供的WebSocket扩展

首先引入Tomcat依赖，如果是Spring Boot方式，可以直接引入Spring Boot Web依赖，因为这个依赖嵌入了Tomcat。

```
<dependency>
    <groupId>org.springframework.boot</groupId>
    <artifactId>spring-boot-starter-web</artifactId>
</dependency>
```

然后创建WebSocket服务端。将一个类定义为WebSocket服务端，只需加上@ServerEndpoint注解。这个注解的源码如下：

```
@Retention(RetentionPolicy.RUNTIME)
@Target(ElementType.TYPE)
public @interface ServerEndpoint {
    String value();
    String[] subprotocols() default {};
    Class<? extends Decoder>[] decoders() default {};
    Class<? extends Encoder>[] encoders() default {};
    public Class<? extends ServerEndpointConfig.Configurator> configurator()
        default ServerEndpointConfig.Configurator.class;
}
```

参数说明：

- value：用来定义服务端连接的路径。
- encoders：编码器。WebSocket API提供了encoders和decoders用于 WebSocket Message 与传统Java类型之间的转换。编码器输入Java对象，生成一种表现形式，这种表现形式能够被转换成WebSocket Message，用于发送请求的时候发送Object对象，实则是JSON数据。
- decoders：解码器。解码器执行与编码器相反的动作，它读入WebSocket消息，然后输出Java对象。

定义完成后，得到一个服务端类。代码如下：

```
@Component
@ServerEndpoint("/{userId}")
```

```java
public class WebSocketServer {
    //连接成功时调用
    @OnOpen
    public void onOpen(Session session, @PathParam("userId") Long userId){
        System.out.println("WebSocketServer 收到连接: " + session.getId() + ",当前用户: " + userId);
    }
    //收到客户端信息
    @OnMessage
    public void onMessage(Session session, String message) throws IOException {
        message = "WebSocketServer 收到连接: " + session.getId() + ",已收到消息: " + message;
        session.getBasicRemote().sendText(message);
    }
    //连接关闭
    @OnClose
    public void onclose(Session session){
        System.out.println("连接关闭");
    }
    //发生错误时的处理方法
    @OnError
    public void onError(Session session, Throwable error) {
        log.error("发生错误,原因:"+error.getMessage());
        error.printStackTrace();
    }
}
```

这里提供了四个方法，分别在不同的阶段执行。

- @OnOpen：服务连接时执行，在这里可以获得连接的初始信息，做一些数据初始化的操作，例如统计在线人数等。
- @OnClose：服务连接关闭时执行，做一些数据清除工作，例如提醒用户下线、清除缓存、统计人数等。
- @OnMessage：收到客户端发送的消息时执行该方法。在这里就可以将客户端的状态同步到其他客户端上。该方法接收的message消息类型除了文本类型String之外，还可以是二进制类型，即byte[] message。
- @OnError：发生错误时调用的方法。在该方法中，可以先将错误捕捉打印到日志，而不影响WebSocket继续向下执行。

以上就是后端服务使用Javax包的类提供WebSocket服务端的过程。前端使用HTML5支持的标准的WebSocket即可。

（2）Spring实现的标准WebSocket

Spring提供的WebSocket标准在实现封装上与Javax包提供的类比较类似，但使用起来更简便，并且支持的消息类型更多。

首先依然是引入依赖。

```xml
<dependency>
    <groupId>org.springframework.boot</groupId>
    <artifactId>spring-boot-starter-websocket</artifactId>
</dependency>
```

然后定义一个handle类，处理WebSocket服务的状态。代码如下：

```java
@Component
public class SpringSocketHandle implements WebSocketHandler {
//连接建立
 @Override
 public void afterConnectionEstablished(WebSocketSession session) throws Exception {
     System.out.println("SpringSocketHandle, 收到新的连接: " + session.getId());
 }
//收到消息
 @Override
 public void handleMessage(WebSocketSession session, WebSocketMessage<?> message) throws Exception {
     String msg = "SpringSocketHandle, 连接: " + session.getId() + ", 已收到消息。";
     System.out.println(msg);
     session.sendMessage(new TextMessage(msg));
 }
//发生错误
 @Override
 public void handleTransportError(WebSocketSession session, Throwable exception) throws Exception {
     System.out.println("WS 连接发生错误");
 }
//连接关闭
 @Override
 public void afterConnectionClosed(WebSocketSession session, CloseStatus closeStatus) throws Exception {
     System.out.println("WS 关闭连接");
 }
```

参数说明：

- afterConnectionEstablished：连接成功后调用，对应@OnOpen。
- handleMessage：处理发送来的消息，对应@OnMessage。
- handleTransportError：WebSocket连接出错时调用，对应@OnError。
- afterConnectionClosed：连接关闭后调用，对应@OnClose。

其中handleMessage是比较重要的部分，它是消息处理逻辑的中心。此处涉及WebSocketMessage接口，这是一个基础的消息类接口。上层实现了一个带泛型的抽象类AbstractWebSocketMessage<T>。然后以此为基础，将泛型具体化，定义了四个消息相关的子类，代表Spring WebSocket支持的四种消息类型。如下所示。

- TextMessage：文本消息。
- BinaryMessage：二进制消息。

- PingMessage：ping消息，一般用于长连接中的心跳发射，由服务器发送给客户端。
- PongMessage：pong消息，一般是用于对ping消息的响应，也可单方面发送，用于心跳发射。当接收到单方面用于心跳的pong消息，服务端无须做任何回复。ping、pong的操作对应的是WebSocket的两个控制帧，opcode分别是0x9、0xA。

整个WebSocket消息相关类的结构如图7.1所示。

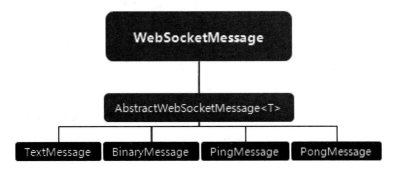

图7.1　WebSocket消息类

然后在Spring中设置开启WebSocket以及配置WebSocket的URL。代码如下：

```
@Configuration
@EnableWebSocket //开启WebSocket
public class SpringSocketConfig implements WebSocketConfigurer {
    //注入之前自定义的Handle处理类
    @Autowired
    private SpringSocketHandle springSocketHandle;
    //注册处理类，并绑定URL，设置跨域支持
    @Override
    public void registerWebSocketHandlers(WebSocketHandlerRegistry registry) {
        registry.addHandler(springSocketHandle,
"/my-ws").setAllowedOrigins("*");
    }
}
```

以上就是在Spring中使用标准的WebSocket的步骤。前端使用HTML5支持的标准的WebSocket即可。

（3）Spring实现的STOMP封装组件

STOMP（Simple Text-Orientated Messaging Protocol）是面向消息的简单文本协议。同HTTP在TCP套接字上添加请求一响应模型层一样，STOMP 在 WebSocket 之上提供了一个基于帧的线路格式层，用来定义消息语义。

首先依然是导入依赖，和Spring的标准WebSocket所使用的依赖相同。

然后配置WebSocketConfig，代码如下：

```
@Configuration
@EnableWebSocketMessageBroker
public class WebSocketConfig implements WebSocketMessageBrokerConfigurer {
    @Override
    public void configureMessageBroker(MessageBrokerRegistry config){
```

```
        // 注册URL
        config.enableSimpleBroker("/broadcast","/user","/alone","/many");
    }
    @Override
    public void registerStompEndpoints(StompEndpointRegistry registry) {
        //注册两个STOMP的endpoint,分别用于广播和点对点
        //广播类型
        registry.addEndpoint("/publicServer"). setAllowedOriginPatterns("*").
withSockJS();
        //点对点类型
        registry.addEndpoint("/privateServer").setAllowedOriginPatterns("*").
withSockJS();
    }
}
```

然后创建一个Controller,使用Spring Boot封装好的SimpMessagingTemplate来调用convertAndSendToUser()实现往指定的userId发送消息。代码如下:

```
@Controller
@Slf4j
public class ChatRoomController {
    @Autowired
    public SimpMessagingTemplate template;
    @GetMapping("/chatroom")
    public String  chatroom (){
        return "chatroom";          //聊天室页面
    }
    //群发
    //SendTo 发送至 Broker 下的指定订阅路径
    @SendTo("/many")
    public ChatRoomResponse mass(ChatRoomRequest chatRoomRequest){
        //群发测试
        log.info("name = " + chatRoomRequest.getName()+";chatValue = " +
chatRoomRequest.getChatValue());
        ChatRoomResponse response=new ChatRoomResponse();
        response.setName(chatRoomRequest.getName());
        response.setChatValue(chatRoomRequest.getChatValue());
        return response;
    }
    //点对点,私聊
    @MessageMapping("/alone")
    public ChatRoomResponse alone(ChatRoomRequest chatRoomRequest){
        //一对一测试
        log.info("userId = " + chatRoomRequest.getUserId());
        log.info("name = " + chatRoomRequest.getName());
        log.info("chatValue = " + chatRoomRequest.getChatValue());
        ChatRoomResponse response=new ChatRoomResponse();
        response.setName(chatRoomRequest.getName());
        response.setChatValue(chatRoomRequest.getChatValue());
        this.template.convertAndSendToUser(chatRoomRequest.getUserId()+"",
"/alone",response);
        return response;
    }
}
```

这里的ChatRoomRequest和ChatRoomResponse都是自定义的类，可以根据具体要求定义属性。这里的属性key与前端使用的key能够对应就可以了。

前端需要借助Stomp.js和sockjs-client来进行WebSocket操作。

创建连接对象后，订阅后端注册的URL地址，代码如下：

```
//建立连接对象（还未发起连接）
var socket=new SockJS("/my-ws");
//获取 STOMP 子协议的客户端对象
var stompClient = Stomp.over(socket);
//订阅群发地址
stompClient.subscribe('/many',function(response){
    ...
}
//订阅单发地址
stompClient.subscribe('/user/' + userId + '/alone',function(response){
    ...
}
```

订阅之后，就可以发送消息了，代码如下：

```
//群发
stompClient.send("/many",{},JSON.stringify(postValue));
//单独发送
stompClient.send("/alone",{},JSON.stringify(postValue));
```

以上就是使用Spring封装的STOMP组件来实现前后端WebSocket协议的过程。

（4）Netty实现WebSocket

Netty 是一个基于NIO的客户服务器端框架，使用Netty能够快速、方便地开发出网络应用，它极大地简化了例如TCP或UDP Socket服务端的开发。Netty融合了多种协议的实现，例如FTP、SMTP、HTTP等各种二进制文本协议。它的官网地址为https://netty.io/index.html。

在Netty中使用WebSocket，首先要引入如下依赖：

```xml
<dependency>
    <groupId>io.netty</groupId>
    <artifactId>netty-all</artifactId>
    <version>4.1.78.Final</version>
</dependency>
```

接下来创建一个处理器，这个处理器用来处理服务端与客户端交互的各个状态。例如连接、断开连接、服务端接收到消息等。在这里我们可以区分消息类型做出不同的处理方式，这个处理器不要自定义，而是继承自SimpleChannelInboundHandler。

```java
@Slf4j
public class MyWebSocketServerHandler extends SimpleChannelInboundHandler<Object>{
    //服务端用于WebSocket握手的类
    private WebSocketServerHandshaker handshaker;
    @Override
    protected void channelRead0(ChannelHandlerContext ctx, Object msg) throws Exception {
        if (msg instanceof FullHttpRequest) {
```

```java
                //服务端客户端建立连接时使用，以http请求形式接入，但是是用来告诉服务端要将协议升级为WebSocket
                handleHttpRequest(ctx, (FullHttpRequest) msg);
            } else if (msg instanceof WebSocketFrame) {
                //处理websocket客户端的消息
                handlerWebSocketFrame(ctx, (WebSocketFrame) msg);
            }
        }
        @Override
        public void channelActive(ChannelHandlerContext ctx) throws Exception {
            //添加连接
            ChannelMessageUtil.addChannel(ctx.channel());
        }
        @Override
        public void channelInactive(ChannelHandlerContext ctx) throws Exception {
            //断开连接
            ChannelMessageUtil.removeChannel(ctx.channel());
        }
        @Override
        public void channelReadComplete(ChannelHandlerContext ctx) throws Exception {
            ctx.flush();
        }
        private void handlerWebSocketFrame(ChannelHandlerContext ctx, WebSocketFrame frame) {
            // 判断请求是否为关闭连接
            if (frame instanceof CloseWebSocketFrame) {
                handshaker.close(ctx.channel(),(CloseWebSocketFrame)frame.retain());
                return;
            }
            // 判断是否ping消息
            if (frame instanceof PingWebSocketFrame) {
                ctx.channel().write(new PongWebSocketFrame(frame.content().retain()));
                return;
            }
            // 判断是不是文本消息
            if (!(frame instanceof TextWebSocketFrame)) {
                throw new UnsupportedOperationException(String.format("%s frame types not supported", frame.getClass().getName()));
            }
            // 把服务器收到的消息发送到通道
            String request = ((TextWebSocketFrame) frame).text();
            log.info("服务端收到：" + request);
            //封装消息
            JSONObject msg = JSONUtil.createObj()
                    .set("date", DateUtil.now())
                    .set("msg", request);
            TextWebSocketFrame tws = new TextWebSocketFrame(msg.toString());
            // 群发,排除掉当前的channelId
            ChannelMessageUtil.send2All(tws, ctx.channel().id());
        }
        /**
         * 特殊的HTTP请求，创建websocket时处理
         */
```

```
    private void handleHttpRequest(ChannelHandlerContext ctx, FullHttpRequest req) {
        //要求Upgrade为websocket,过滤掉get/Post
        if (!req.decoderResult().isSuccess() ||
(!"websocket".equals(req.headers().get("Upgrade")))) {
            //若不是websocket方式,则创建BAD_REQUEST的req,返回给客户端
            sendHttpResponse(ctx, req, new DefaultFullHttpResponse
(HttpVersion.HTTP_1_1, HttpResponseStatus.BAD_REQUEST));
            return;
        }
        //创建握手
        handshaker = new WebSocketServerHandshaker13
("ws://127.0.0.1:8088/websocket", null, true, 1024 * 1024 * 6);
        if (handshaker == null) {
            //握手失败
            WebSocketServerHandshakerFactory.sendUnsupportedVersionResponse
(ctx.channel());
        } else {
            //返回握手信息
            handshaker.handshake(ctx.channel(), req);
        }
    }
    /**
     * 拒绝不合法的请求,并返回错误信息
     */
    private void sendHttpResponse(ChannelHandlerContext ctx, FullHttpRequest req,
DefaultFullHttpResponse res) {
        // 当前回应码是否为200
        if (res.status().code() != 200) {
            ByteBuf buf = Unpooled.copiedBuffer(res.status().toString(),
CharsetUtil.UTF_8);
            res.content().writeBytes(buf);
            buf.release();
        }
        ctx.channel().writeAndFlush(res);
    }
}
```

其中ChannelMessageUtil是自定义的通道处理工具。通过它可以添加通道、移除通道、查找通道、群发消息等。

完成以上内容后,就可以在Spring Boot启动类中添加Netty服务了。

```
@SpringBootApplication
public class Application {
    public static void main(String[] args) {
        ApplicationContext applicationContext = SpringApplication.run
(Application.class, args);
        //从ioc容器获取NioWebSocketServer并启动websocket服务器
        MyWebSocketServer myWebSocketServer = applicationContext.getBean
("myWebSocketServer",MyWebSocketServer.class);
        myWebSocketServer.run();
    }
}
```

至此,通过Netty实现了WebSocket服务端的创建。前端使用标准的WebSocket即可。

7.1.2 框架搭建

前面讲述了在Java中使用WebSocket协议的四种方式,本章项目选取的模拟聊天室使用Javax包提供的WebSocket包来实现,这种方式最基础。使用这种方式可以更加清晰地了解WebSocket的原理。掌握了这种方式之后,再循序渐进地进行其他方式的学习和使用,印象会更深刻。以下为该项目的技术框架部分:

- 开发框架:Spring Boot。
- 数据库:MySQL。
- 后台框架:Spring、Spring MVC、MyBatis。
- 前端框架:LayUI。
- 通讯协议:WebSocket。

除了WebSocket外,其他部分的整合在前面的章节中已经详细描述了。本章重点讲述如何在Spring Boot中整合WebSocket。

在Spring Boot中整合WebSocket非常简单,首先引入对应的依赖:

```xml
<dependency>
    <groupId>org.springframework.boot</groupId>
    <artifactId>spring-boot-starter-websocket</artifactId>
</dependency>
```

然后定义一个WebSocket处理器,继承自WebSocketHandler,处理器中重写连接建立、连接关闭、消息处理、异常处理等方法。然后加上@Component注解,使Spring能够自动扫描组件。

```java
@Component
public class WebSocketServer implements WebSocketHandler {
    /**
     * 连接建立时调用
     * @param webSocketSession
     * @throws Exception
     */
    @Override
    public void afterConnectionEstablished(WebSocketSession webSocketSession) throws Exception {
        ...
    }
    /**
     * 处理收到的消息
     * @param webSocketSession
     * @param webSocketMessage
     * @throws Exception
     */
    @Override
    public void handleMessage(WebSocketSession webSocketSession, WebSocketMessage<?> webSocketMessage) throws Exception {
        ...
```

```
        }
        /**
         * 发生错误时调用
         * @param webSocketSession
         * @param throwable
         * @throws Exception
         */
        @Override
        public void handleTransportError(WebSocketSession webSocketSession, Throwable throwable) throws Exception {
            ...
        }
        /**
         * 连接关闭时调用
         * @param webSocketSession
         * @param closeStatus
         * @throws Exception
         */
        @Override
        public void afterConnectionClosed(WebSocketSession webSocketSession, CloseStatus closeStatus) throws Exception {
            ...
        }
    }
```

然后创建一个配置类，继承自WebSocketConfigurer，在这里重写注册处理器的方法，将刚才定义的处理器注入进来，设置好连接路径的匹配规则和跨域，最后加上@Configuration和@EnableWebSocket注解。@Configuration注解用来表示这是一个配置类，@EnableWebSocket注解表示当前服务要开启WebSocket协议支持。

```
@Configuration
@EnableWebSocket
public class WebSocketConfig implements WebSocketConfigurer {
    //注入之前自定义的Handle处理类
    @Autowired
    private WebSocketServer webSocketServer;
    //注册处理类并绑定URL，设置跨域支持
    @Override
    public void registerWebSocketHandlers(WebSocketHandlerRegistry registry){
        registry.addHandler(webSocketServer,
"/webSocket/*").setAllowedOrigins("*");
    }
}
```

经过以上配置之后，WebSocket与Spring Boot的融合就已完成。接下来我们可以配置数据库以及其他框架，配置完成后，Spring的配置文件大致如下：

```
#yml
server:
  port: 8088 #服务端口
spring: #数据源
  datasource:
    ssm:
```

```
            driver-class-name: com.mysql.cj.jdbc.Driver #数据源驱动器
            password: 123456 #数据库密码
            url: jdbc:mysql://localhost:3306/chat?serverTimezone=GMT%2B8 #数据库连接URL
            username: root #数据库用户名
```

以上内容完成后，就完成了框架的初步搭建，接下来就可以进行功能实现了。

7.2 项目前期准备

7.2.1 项目需求说明

模拟聊天室，主要用来实现用户的上线、下线、群发消息、给指定用户发送消息等功能。用户登录后，可以获取当前在线用户，并且将自己的上线消息同步广播到其他用户。成功上线之后，能够群发或者单独发送消息内容。下线时，能够将下线状态同步广播到其他用户。

7.2.2 系统功能设计

根据模拟聊天室要实现的主要功能，可以得到该系统的系统功能结构，如图7.2所示。

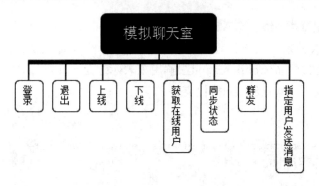

图7.2 模拟聊天室功能结构

- 登录：用户输入用户名和密码，验证登录，成功后跳转至聊天室，失败则跳转至错误页。
- 上线：与服务器端建立连接，广播上线消息。
- 下线：与服务器端断开连接，广播下线消息。
- 获取在线用户：将在线用户展示在页面上，用于选择聊天对象。
- 群发：将消息群发给当前在线的其他用户。
- 指定用户发送消息：选定某个用户，单独发送消息。

7.2.3 系统数据库设计

这个模拟聊天室的设计非常简单，主要目的是使用Spring封装WebSocket组件，所以对应的数据库设计也非常简单，只涉及一个用户表（member），包含的主要信息有用户名、密码、最后登录时间等。以下是用户表的建表语句。

```
    DROP TABLE IF EXISTS 'member';
    CREATE TABLE 'member' (
      'member_id' tinyint unsigned NOT NULL,
      'address_id' smallint unsigned NOT NULL,
      'email' varchar(50) CHARACTER SET utf8 COLLATE utf8_general_ci NULL DEFAULT NULL,
      'username' varchar(16) CHARACTER SET utf8 COLLATE utf8_general_ci NOT NULL,
      'password' varchar(40) CHARACTER SET utf8 COLLATE utf8_bin NULL DEFAULT NULL,
      'last_login_time' timestamp(0) NOT NULL DEFAULT CURRENT_TIMESTAMP ON UPDATE CURRENT_TIMESTAMP(0),
      PRIMARY KEY ('member_id') USING BTREE,
      INDEX 'idx_fk_address_id'('address_id') USING BTREE
    ) ENGINE = InnoDB AUTO_INCREMENT = 4 CHARACTER SET = utf8 COLLATE = utf8_bin ROW_FORMAT = Dynamic;
    -- ----------------------------
    -- Records of member
    -- ----------------------------
    INSERT INTO 'member' VALUES (1, 3, '', 'lijun', '123456', '2022-06-24 10:25:12');
    INSERT INTO 'member' VALUES (2, 4, '', 'limei', '123456', '2022-06-24 10:25:14');
    INSERT INTO 'member' VALUES (3, 3, 'DF', 'liuqing', '123456', '2022-06-24 10:25:16');
    SET FOREIGN_KEY_CHECKS = 1;
```

7.2.4 系统文件说明

本项目系统文件的组织结构如图7.3所示。

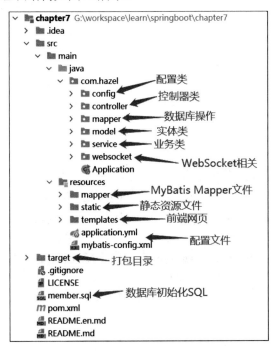

图7.3 模拟聊天室系统文件组织结构

每个部分说明如下：

- config：配置类，数据源和WebSocket基础配置。

- controller：控制器类，用于处理请求，是与前端交互的主要窗口。
- mapper：数据库操作，Dao层接口。
- model：实体类，数据库表对应的Entity实体，以及一些常用的POJO。
- service：业务类接口。
- websocket：WebSocket的主要处理器。
- resources：资源文件，包括MyBatis的mapper文件、静态资源（例如CSS、JS等）、网页模板文件、Spring的配置文件等。

7.3 项目前端设计

7.3.1 登录与退出

登录时，输入用户名和密码，校验通过跳转至聊天室，不通过则弹出提示信息，然后跳转回登录页，如图7.4和图7.5所示。

退出时，点击左上角的退出按钮，页面跳转到登录页面，如图7.6所示。

图7.4　登录页面　　　　　　图7.5　错误提示页面　　　　　　图7.6　退出页面

登录与退出与前面系统类似，调用后台提供的登录与退出接口即可，这里不再详述。

7.3.2 聊天室主页面

模拟聊天室的前端设计主要集中在聊天室这个主页面上，其中包括当前登录用户、当前在线用户、上下线动态、消息窗口、发送工具等，如图7.7所示。

用户登录成功后，首先获取当前在线用户列表，并显示到页面中。

```
//获取当前在线用户列表
$.get("/onlineusers?currentuser="+user,function(data){
    for(let i=0;i<data.length;i++){
        let onlineUser={user:data[i]}
        //push到用户数组
        onlineUsers.push(onlineUser);
         //重载表格
        onlineTable.reload()
    }
})
```

这里用到了LayUI框架中的Table组件重载onlineTable.reload()。

图7.7 聊天室主页面

然后我们可以同时发送WebSocket建立连接的请求,与后端服务器建立连接。

```
//判断当前浏览器是否支持WebSocket
if ('WebSocket' in window) {
    //支持,创建连接
    websocket = new WebSocket(websocketUrl+user);
}
```

如果浏览器支持WebSocket,那么建立标准的WebSocket协议,创建连接。创建连接之后,就可以进行消息监听了,这样在服务器推送消息时,客户端的WebSocket就可以即时收到消息,进行相应的处理。

```
//WebSocket消息监听
websocket.onmessage = function(event) {
    //收到的数据
    var data=JSON.parse(event.data);
    if(data.to==0){//上线消息
        ...//接收到用户上线的一系列处理
    }else if(data.to==-2){//下线消息
    ...//接收到用户下线的一系列处理
    else {
        ...//处理接收到的消息内容
    }
}
```

7.3.3 群发消息

群发消息时,在左下角输入消息内容,点击发送按钮进行发送。如果消息内容为空,不能正常发送,则给出提示,如图7.8所示。

群发消息后,当前在线的所有用户都能收到该消息,如图7.9所示。

图7.8 群发消息提示

图7.9 群发消息

对应的JavaScript代码如下:

```javascript
//群发消息
$("#broadcastinfo").click(function(){
    //获取消息输入框中的内容
    let msg = $("#msg").val();
    //判断输入内容是否为空
    if(msg == ""){
        layer.msg("请输入消息内容");
        return;
    }
    //构造消息对象,与后台Message对象属性一致
    let data = {
        from:"系统消息",
        to:-1,
        text:msg
    };
    //发送消息
    websocket.send(JSON.stringify(data));
});
```

其中#broadcastinfo为群发按钮的id,这段代码为它绑定click事件,当点击该按钮时,获取输入框中的内容。如果输入框内容为空,使用Layer组件给出弹窗提示。如果消息内容不为空,构造消息对象,该对象属性与后台Message对象的属性一致,以便解析消息对象。然后就可以发送消息了,发送之后,在线用户都会收到该消息。

7.3.4 给指定用户单独发送消息

单独发送消息的输入框与按钮位于消息主窗口下方。发送之前需要在左侧在线用户列表中选择一个用户。如果未选择用户,则点击发送按钮会给出提示,如图7.10所示。

图7.10 发送提示

当选择用户后,如果消息内容为空也会给出提示,参考群发消息部分。如果消息内容正常发送,在发送用户与接收用户的消息窗体中会显示该条信息,如图7.11和图7.12所示。

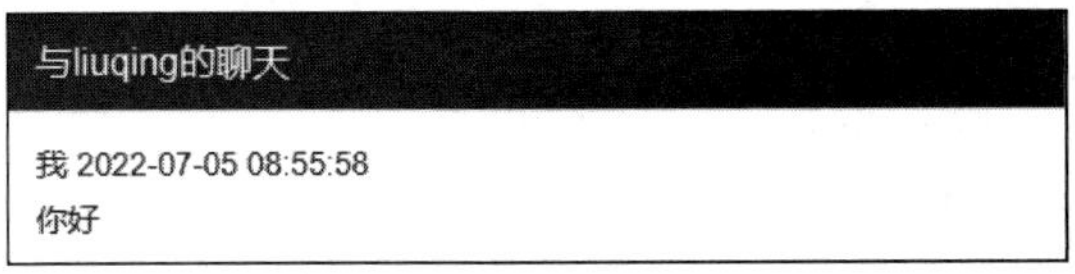

图7.11　发送者聊天窗口 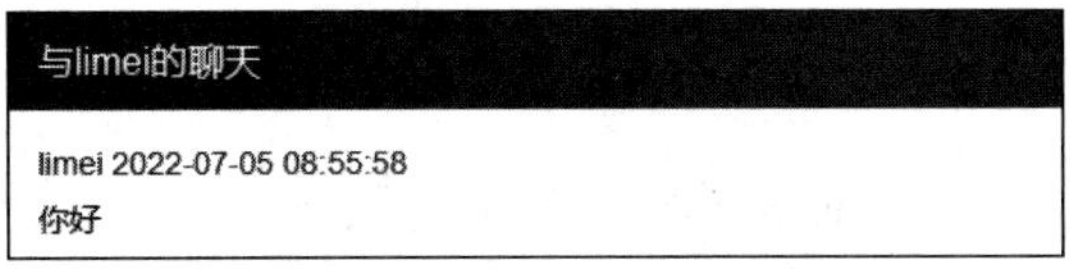　　　　　　　图7.12　接收者聊天窗口

这部分对应的JavaScript代码如下：

```javascript
//给指定用户发送消息
$("#send").click(function() {
    //判断是否选择了目标用户
    if ($("body").data("to")==undefined) {
        layer.msg("请选择聊天对象");
        return false;
    }
    //获取消息内容
    let msg = $("#myinfo").val();
    //消息内容为空
    if(msg == ""){
        layer.msg("请输入消息内容");
        return;
    }
    //构造消息对象，与后台的Message对象属性一致
    let toUser = $("body").data("to");
    let data = {
        from:user,
        to:toUser,
        text:msg
    }
    //发送消息
    websocket.send(JSON.stringify(data));
    //将消息同步到自己的消息窗口中，并下拉滚动
    let formatDate = $.format.date(new Date(),"yyyy-MM-dd HH:mm:ss");
    $("#info-container").append("<div class='bg-success'><label class='text-info'>我 " + formatDate + "</label><div class='text-info'>" + $("#myinfo").val() + "</div></div><br>");
    scrollToBottom("info");
    //清空输入框
    $("#myinfo").val("");
});
```

其中#send是发送按钮的id，为它绑定click事件。成功发送消息后需要将消息同步到发送者的消息窗口，因为发送者自己不会收到自己发送的消息内容。同步后，如果消息多到已经超出窗体，则需要将窗体滚动下拉到显示该条信息。最后清空输入框。

选择聊天对象用户时，需要借助LayUI框架中的Table组件，为它绑定行单击事件，代码如下：

```javascript
//触发行单击事件
table.on('row(users)', function(obj){
    let chooseUser = obj.data;
    $("#title").text("与"+chooseUser.user+"的聊天");
```

```
            $("body").data("to",chooseUser.user);
    });
```

其中users是给表格指定的过滤标识，row(users)表示表格过滤标识为users的行元素。选中用户后，更新聊天窗体的标题，并将用户设置到页面数据中，便于发送消息时调用。

7.3.5 上线与下线

用户登录时伴随用户上线,用户退出系统或者页面发生不可知错误时伴随用户下线。用户上下线时，需要将其状态同步通知到其他用户，如图7.13所示。

图7.13 上下线动态

上线即建立WebSocket连接，参考聊天室主页面部分。当前页面关闭或跳转到其他页面后，WebSocket就可以同时收到关闭消息，即下线。当收到建立连接与关闭连接的消息后，后台会向前端广播这条消息，所有在线的其他用户都会收到该条消息，接着更新上下线动态，同时更新在线用户列表。对应的代码如下：

```
    if(data.to==0){//上线消息
        if(data.text!=user)
        {
            //增加一条用户信息
            let formatDate = $.format.date(new Date(),"yyyy-MM-dd HH:mm:ss");
            let newOnlineUser = {
                user:data.text
            }
            onlineUsers.push(newOnlineUser)
            //重载表格
            onlineTable.reload();
            //更新上下线动态
            let newMemeber = "<div><h5 style='color:darkgreen;font-weight: bold;'>"
+formatDate+"</h5><text>"+data.text + "上线了"+"</text></div>";
            $("#member-container").append(newMemeber)
            scrollToBottom("member");
        }
    }else if(data.to==-2){//下线消息
        if(data.text!=user)
        { //删除列表中该条用户
            $.each(onlineUsers, function (i, item) {
                if(item != null && item.user != null && item.user != '' && item.user ==
data.text){
                    onlineUsers.splice(i,1);
                }
            })
            //重载表格
            onlineTable.reload();
            //更新上下线动态
            let formatDate = $.format.date(new Date(),"yyyy-MM-dd HH:mm:ss");
            let outMemeber = "<div><h5 style='color:darkred;font-weight: bold;'>"
+formatDate+"</h5><text>"+data.text + "下线了"+"</text></div>";
            $("#member-container").append(outMemeber)
```

```
            scrollToBottom("member");
        }
    }
```

消息对象中的to属性标识消息的目标用户,我们可以为其规定一些特殊的值,表示一些特殊的消息。例如本例中,其值为0代表用户上线,为-2代表用户下线。

以上就是前端页面的设计。读者可以参考目前市面上常用的聊天软件,对本章示例进行扩展,例如增加好友与群的功能、设置用户头像等。

7.4 项目后端实现

7.4.1 上线与下线

上线与下线对应WebSocket服务中的连接建立与关闭,在Spring提供的WebSocket包中,对应的方法为afterConnectionEstablished与afterConnectionClosed。创建的自定义处理器继承自WebSocketHandler,重写这两个方法即可。

用户上线,建立连接后,需要同步广播其上线状态,更新目前在线用户列表,代码如下:

```
/**
 * 连接建立时调用
 * @param webSocketSession
 * @throws Exception
 */
@Override
public void afterConnectionEstablished(WebSocketSession webSocketSession) throws Exception {
    //获取当前登录用户名
    String userName = getUserName(webSocketSession);
    //将用户的session对象放入map
    sessionPools.put(userName, webSocketSession);
    //人数加1
    addOnlineCount();
    // 向其他用户广播上线消息
    Message msg = new Message();
    msg.setDate(new Date());
    msg.setTo("0");
    msg.setText(userName);
    //构造消息对象
    TextMessage message = new TextMessage(JSON.toJSONString(msg,true));
    //发送广播
    broadcast(message);
}
```

用户建立连接后,首先更新连接数量。addOnlineCount()为更新连接数的方法:

```
//静态变量,用来记录当前在线连接数。应该把它设计成线程安全的
private static AtomicInteger onlineNum = new AtomicInteger();
//人数增加1
```

```java
public static void addOnlineCount(){
    onlineNum.incrementAndGet();
}
```

这里用到了AtomicInteger，AtomicInteger类是系统底层保护的int类型，通过线程安全的方式进行int类型的加、减操作。AtomicInteger通过调用构造函数可以直接创建。为了线程安全，这里我们使用该类作为计数器。

用作计数器时，AtomicInteger提供了一些以原子方式执行加法和减法操作的方法，如表7.2所示。我们这里用到的是incrementAndGet方法。

表7.2　AtomicInteger计数器方法

方 法 名	方法描述
addAndGet	以原子方式将给定值添加到当前值，并在添加后返回新值
getAndAdd	以原子方式将给定值添加到当前值并返回旧值
incrementAndGet	以原子方式将当前值递增1并在递增后返回新值。它相当于i++操作
getAndIncrement	以原子方式递增当前值并返回旧值。它相当于++i操作
decrementAndGet	原子地将当前值减1并在减量后返回新值。它等同于i--操作
getAndDecrement	以原子方式递减当前值并返回旧值。它相当于--i操作

连接建立后，就可以向其他用户广播上线消息。首先构造消息对象，前面提到Spring提供的WebSocket包支持四种消息类型，这里我们使用简单的TextMessage文本消息。广播群发的实现内容如下：

```java
// 群发消息
public void broadcast(WebSocketMessage message){
    //获取当前所有连接，遍历循环
    for (WebSocketSession session: sessionPools.values()) {
        try {
            //发送消息
            sendMessage(session, message);
        } catch(Exception e){
            e.printStackTrace();
            continue;
        }
    }
}
//发送消息
public void sendMessage(WebSocketSession session, WebSocketMessage message) throws IOException {
    if(session != null){
        synchronized (session) {
            System.out.println("发送数据：" + message);
            session.sendMessage(message);
        }
    }
}
```

群发消息的底层还是调用单独的session发送消息，只不过在调用前增加一个批量操作，就可以实现群发效果。在缓存里拿到目前所有在线连接的session，然后遍历循环sessionPools.values()。

sessionPools是我们定义用来存放连接的集合，这里为了线程安全使用了ConcurrentHashMap。

```
//concurrent包的线程安全设置，用来存放每个客户端对应的WebSocketSession对象，目前仅支持一
个账号登录一个客户端
    private static ConcurrentHashMap<String, WebSocketSession> sessionPools = new
ConcurrentHashMap<>();
```

从JDK1.8开始，ConcurrentHashMap数据结构与HashMap保持一致，两者均为数组加链表加红黑树，通过乐观锁和Synchronized来保证线程安全。当多线程并发向同一个散列桶添加元素时，若散列桶为空，此时触发乐观锁机制，线程会获取到桶中的版本号，在添加节点之前，判断线程中获取的版本号与桶中实际存在的版本号是否一致，若一致，则添加成功，若不一致，则让线程自旋。若散列桶不为空，此时使用Synchronized来保证线程安全，先访问到的线程会给桶中的头节点加锁，从而保证线程安全。

发送消息时，执行WebSocketSession接口提供的sendMessage方法。由于WebSocketSession是一个接口，所以要有实现类来实现对应的方法。从Spring 4.0版本开始，默认提供StandardWebSocketSession作为它的实现类。但如果我们分析它的源码会发现，它并没有sendMessage方法，原因是Spring规定发送消息时不能使用同步的方式，必须使用异步方式来发送消息，所以这里会调用ConcurrentWebSocketSessionDecorator类中的sendMessage方法，将要发送的信息放置在队列，然后执行算法分析是否能够发送消息。

```
//Spring官方源码
@Override
public void sendMessage(WebSocketMessage<?> message) throws IOException {
    //判断是否能够发送，两个标准：1是否超出数量限制，2进程是否正常
    if (shouldNotSend()) {
        return;
    }
    //能够发送，加入缓存队列
    this.buffer.add(message);
    this.bufferSize.addAndGet(message.getPayloadLength());
    do {
        if (!tryFlushMessageBuffer()) {
            if (logger.isTraceEnabled()) {
                logger.trace(String.format("Another send already in progress: " +
                    "session id '%s':, \"in-progress\" send time %d (ms), buffer size %d bytes",
                    getId(), getTimeSinceSendStarted(), getBufferSize()));
            }
            checkSessionLimits();
            break;
        }
    }
    while (!this.buffer.isEmpty() && !shouldNotSend());
}
```

如果能够发送，就会执行StandardWebSocketSession中对应的发送方法。

```
//发送文本消息
@Override
```

```java
    protected void sendTextMessage(TextMessage message) throws IOException {
        getNativeSession().getBasicRemote().sendText(message.getPayload(),
message.isLast());
    }
```

这样就完成了用户上线消息的推送。下线消息同理,只不过触发的方法为afterConnectionClosed。

```java
/**
 * 连接关闭时调用
 * @param webSocketSession
 * @param closeStatus
 * @throws Exception
 */
@Override
public void afterConnectionClosed(WebSocketSession webSocketSession, CloseStatus closeStatus) throws Exception {
    // 获取当前登录用户名
    String userName = getUserName(webSocketSession);
    // 将用户对应的session从map中移除
    sessionPools.remove(userName);
    // 用户数减1
    subOnlineCount();
    // 向其他用户广播下线消息
    Message msg = new Message();
    msg.setDate(new Date());
    msg.setTo("-2");
    msg.setText(userName);
    //构造消息对象
    TextMessage message = new TextMessage(JSON.toJSONString(msg,true));
    //发送广播
    broadcast(message);
}
```

7.4.2 发送消息

发送消息分为群发消息和给指定用户单独发送消息。其实两者原理一样,群发消息可以理解为批量的单独发送。接收到用户要发送的消息时,触发的方法为handleMessage,这也是比较核心的方法。由于我们使用的是Spring实现的包,所以这里要使用它所支持的消息类型,即实现了WebSocketMessage接口的消息类型。所以需要做转换。

```java
/**
 * 处理收到的消息
 * @param webSocketSession
 * @param webSocketMessage
 * @throws Exception
 */
@Override
public void handleMessage(WebSocketSession webSocketSession, WebSocketMessage<?> webSocketMessage) throws Exception {
    //收到客户端信息后,根据接收人的username把消息推下去或者群发
    // to=-1群发消息
```

```java
        //目前仅处理文本消息
        if(webSocketMessage instanceof TextMessage){
            Message msg=JSON.parseObject(webSocketMessage.getPayload().toString(),
Message.class);
            msg.setDate(new Date());
            //构造消息对象
            TextMessage message = new TextMessage(JSONObject.toJSONString(msg));
            if (msg.getTo().equals("-1")) {
                //群发
                broadcast(message);
            } else {
                //发送给指定用户
                sendInfo(msg.getTo(), message);
            }
        }
    }
```

如果消息对象中的to属性被设置为–1，代表这是一条群发消息。否则就是一条对指定用户单独发送的消息。在WebSocketMessage类中，消息内容被放置在payload属性中。

```java
public interface WebSocketMessage<T> {
    T getPayload();//获取消息内容
    int getPayloadLength();//获取消息长度
    boolean isLast();//分片消息使用
}
```

7.4.3 获取当前在线用户列表

获取当前在线用户列表，这里其实获取的是当前连接，借助于WebSocketServer中设置的sessionPools，将连接对应的用户转换成结果集返回即可。

```java
/**
 * 获取当前登录的用户列表
 * @param currentuser
 * @return
 */
@GetMapping("/onlineusers")
public Set<String> onlineUsers(@RequestParam("currentuser") String currentuser) {
    //获取当前WebSocketServer处理器中的session存放map
    ConcurrentHashMap<String, WebSocketSession> map =
WebSocketServer.getSessionPools();
    //遍历map
    Set<String> set = map.keySet();
    Iterator<String> it = set.iterator();
    //返回的结果集
    Set<String> onlineUsers = new HashSet<String>();
    while (it.hasNext()) {
        String entry = it.next();
        if (!entry.equals(currentuser))
            onlineUsers.add(entry);
    }
```

```
    return onlineUsers;
}
```

7.5 项目总结

本章实现了一个模拟聊天室系统,后端使用的主要技术有Spring、Spring MVC、MyBatis和WebSocket框架,前端使用的框架为LayUI,数据库使用的是MySQL。本章读者要掌握的重点是WebSocket协议。知道如何通过Spring Boot整合这些框架并做相应配置,以及如何在实际开发过程中使用这个框架。学习时可参考各个框架官方提供的API手册,同时要回顾前面章节的相关内容,以便理解和记忆。

本章的重点技术:

- **WebSocket协议**:WebSocket概述和原理,熟悉其常用场景。
- **WebSocket服务端**:在Java中采取多种方式使用WebSocket协议,重点是使用Spring提供的标准WebSocket封装组件实现服务端WebSocket。学会本章的使用方式之后,可以将这个项目改造成使用其他方式来实现服务端。
- **WebSocket客户端**:在前端采取多种方式使用WebSocket协议。